Astronomers' Observing Guides

Other Titles in This Series

Star Clusters and How to Observe Them
Mark Allison

Saturn and How to Observe it
Julius Benton

Nebulae and How to Observe Them
Steven Coe

The Moon and How to Observe It
Peter Grego

Supernovae and How to Observe Them
Martin Mobberley

Double and Multiple Stars and How to Observe Them
James Mullaney

Galaxies and How to Observe Them
Wolfgang Steinicke & Richard Jakiel

Forthcoming Title in This Series

Total Solar Eclipses and How to Observe Them
Martin Mobberley

James Mullaney, F.R.A.S.

The Herschel Objects
and How to Observe Them

with 90 Illustrations

 Springer

James Mullaney
Rehoboth Beach
Delaware
USA
arcturussj@aol.com

Series Editor:
Dr. Mike Inglis, BSc, MSc, Ph.D.
Fellow of the Royal Astronomical Society
Suffolk County Community College
New York, USA
inglism@sunysuffolk.edu

Library of Congress Control Number: 2007923721

ISBN-13: 978-0-387-68124-5   e-ISBN-13: 978-0-387-68125-2

Printed on acid-free paper.

9 8 7 6 5 4 3 2 1

springer.com

To my adult children – three bright stars in my heavens.

Colleen Mullaney Lenfestey
Christine Mullaney Takacs
James William Mullaney

Preface

Many active amateur astronomers today, having already surveyed the clusters, nebulae, and galaxies contained in the popular Messier and Caldwell catalogs, are seeking new horizons to explore with their telescopes. None better can possibly be found than those discoveries made by the great English astronomer Sir William Herschel in the late 1700s to early 1800s. But rather than just over a hundred objects found in each of the former two listings, Herschel's catalog contains some 2,500 entries. This sheer number of targets has discouraged most observers – however avid they may be about deep-sky observing – from attempting to explore these unsung wonders.

In a letter in the April, 1976, issue of *Sky & Telescope* magazine the author suggested a way to make Herschel's list more attractive to observers. His discoveries were arranged into eight Classes, designated I to VIII (see Chapter 3). Of these, 1,893 lie in Classes II and III – his faint and very faint nebulae. Dropping these and taking only those entries in the remaining classes as a working list results in 615 objects – a much more manageable and realistic number of targets to view. This letter was followed by a full-length article on this concept in the January, 1978, issue of *Astronomy* magazine and (much later) another one in *Sky & Telescope* for September, 1992. As a result of these published pieces, observation of the Herschel objects began to grow in popularity among the stargazing community, and acting upon the author's suggestion an actual Herschel Club was started by the Ancient City Astronomy Club in St. Augustine, Florida. This local effort was eventually adopted on a national level by the Astronomical League, a federation of most of the astronomy clubs in the United States. (See Appendix 1 for more about this and other Herschel Clubs, including a short-lived one dating back to 1958.) Unfortunately, the target list adopted by these organizations contains a total of only 400 entries rather than the full 615 that I had recommended (although some of its founding members are now going after the entire Herschel catalog). Among them are many objects of Classes II and III – which are for the most part anything but exciting at the eyepiece – while a number of real Herschel showpieces are overlooked.

The book you are now holding in your hands is the author's answer to a long-standing need for a work devoted exclusively to the Herschel objects and their observation by amateur astronomers, with emphasis on the 615 objects of Classes I, IV, V, VI, VII, and VIII. We begin by examining something of Sir William's remarkable life and times (including a bit about his famous sister Caroline and son Sir John), his many amazing astronomical discoveries, and his home-made metal-mirrored reflectors ranging in size from around 6 inches in aperture all the way up to 48 inches (the famed "40-Foot" – for a time the largest telescope in the world), and take a look at his catalogs and his various Classes. Following a discussion on observing techniques, we shall then profile some 165 selected showpieces from his catalog suitable for viewing with backyard telescopes ranging from 2- to 14-inches in size (along with a number of fainter objects lying in the same eyepiece fields with them). This constitutes the real heart of this work, and many readers may wish to jump ahead to those chapters immediately. However, knowing something of Herschel's background, the instruments he used to make his discoveries and the nature of each of his classes will add greatly to the ultimate pleasure of viewing those showpieces. We shall also sample some interesting specimens from Classes II and III so observers can get a feel for what these objects look like, and then highlight a number of showpieces that Herschel strangely missed in his "sweeps" of the heavens – plus several of his discoveries that apparently have disappeared from the sky! Various Herschel Clubs will be discussed in Appendix 1, followed by a selected list of Herschel references in Appendix 2 for those desiring to read more about this amazing astronomical family. Finally, rounding out this book in Appendix 3 is a working roster of the entire 615 objects contained in the above-mentioned six classes for those desiring to see all of them as per the author's original suggestion.

And so now dear reader, together let us retrace the glorious pathway in the sky left by this truly great and ardent observer!

James Mullaney
Rehoboth Beach, Delaware, USA
March 2007

Acknowledgments

There are many people in both the professional and amateur astronomical community who have helped to make this book possible. Among them are the editors at both *Sky & Telescope* and *Astronomy* magazines, who have kindly published my various articles and letters over the years on observing the Herschel objects and on the formation of a Herschel Club. Dr. Nicholas Wagman, past director of the Allegheny Observatory in Pittsburgh, kindly allowed me extensive use of its superb 13-inch Fitz-Clark refractor (and occasionally its 30-inch Brashear refractor as well!) for conducting visual surveys of deep-sky wonders, including double stars and the Herschel objects. Special thanks go to Ronald Wiltshire and Peter Hingley of the Royal Astronomical Society in London (of which William Herschel was its first president) for kindly sharing its resources with me, including a copy of its 1962 reprinting of the combined *New General Catalogue of Nebulae and Clusters of Stars* (the *NGC*) and both *Index Catalogues*. The *NGC* contains a complete descriptive listing of all of Sir William's deep-sky discoveries (as well as those of his son, Sir John, and many others). Deep thanks also go to both the RAS and Rose Taylor of the Photo Science Library, London, for kindly supplying images of the three Herschel's and their telescopes. I am especially indebted to Dr. Mike Inglis, FRAS, for taking most of the CCD images of selected Herschel objects illustrating this book.[*] California astroimager Steve Peters has also kindly supplied a number of his personal images of these objects. And my thanks to Charles Feldman, retired IBM engineer, for loading many of the images used in the first few chapters of this book onto CD-ROMs for me. My editors at Springer – Dr. Harry Blom, Christopher Coughlin, and Jenny Wolkowicki in their New York office, Dr. John Watson, FRAS, at their London office, and general series editor Dr. Mike Inglis, himself – have all been most helpful and a sincere pleasure to work with on this, my third volume[**] for this truly world-class publisher. And finally, I would like to thank my dear wife, Sharon McDonald Mullaney, for her continued encouragement and support during the long process of researching and writing this book.

---

[*]Dr. Inglis used 20 cm-, 25 cm-, and 30 cm-aperture Schmidt-Cassegrain telescopes, under a variety of sky conditions ranging from adequate to superb, during 2005–2006. Image reduction utilized MaximIDL, IRAF, and Adobe Photoshop.
[**]The previous two are *Double and Multiple Stars and How to Observe Them* (Springer, 2005), and *A Buyer's and User's Guide to Astronomical Telescopes and Binoculars* (Springer, 2007).

Contents

Part II Exploring The Herschel Showpieces

Contents

Contents

Part I

William Herschel's Life, Telescopes and Catalogs

Chapter 1

Introduction

Who Was Sir William Herschel?

William Herschel was without question the greatest visual observer who ever lived. Variously regarded as the "Father of Observational Astronomy" and the "Father of Sidereal Astronomy," he single-handedly opened the frontiers of deep space to telescopic exploration. In the course of his grand scheme to study what he called the "construction of the heavens," he discovered literally *thousands* of previously unknown double and multiple stars, star clusters, nebulae, and galaxies. Talk about "Going where no man has gone before" (to borrow a line from *Star Trek*)! Although self-taught and so technically an amateur astronomer, he transformed the world of professional astronomy – which at the time had been largely concerned with the solar system and the positions of the stars – and set it on a course that is still under full sail today. (Incidentally, the word "amateur" is derived from the Latin word "amare" which means "to love" – or more precisely, from "amator" which means "one who loves." An amateur astronomer is one who loves the stars. And surely no one loved them more than did Sir William.) (Fig. 1.1)

From Musician to Stargazer

Herschel was born into a musical family in Hanover, Germany, in 1739 and moved to England around 1770. Like others in his family, his early career was that of a musician – in his case, teaching and orchestrating music for the city of Bath. It was while there that he became fascinated with astronomy. (Some would say "obsessed" better describes it, for on occasion he would actually run home during performances to observe between acts! And later, as a full-time astronomer, he typically observed from dusk to dawn.) He set about making his own telescopes beginning with small refractors, but soon abandoned them for a variety of reasons and turned his attention instead to reflectors, constructing entire instruments including their speculum-metal mirrors entirely himself. (The familiar silver-on-glass telescope mirror was not introduced until long after Herschel's death in 1822.) But Herschel not only became the greatest telescope-maker of his time, but was also an observer the caliber of which the world had never seen before. He used these homemade instruments to "sweep" the heavens for unexplored celestial treasure, his initial "review" being undertaken with a "7-foot" Newtonian reflector at a magnification of 227×. (At that time telescopes were

Fig. 1.1. Sir William Herschel at the age of 55, shown holding a drawing of the planet Uranus which he discovered and two of its satellites (which he also discovered). This is a photograph of a famous pastel portrait done by J. Russell, in 1794. Earlier images of Herschel as a young man are very rare and difficult to find. (Fig. 14.1. shows his appearance in his later years.) Yerkes Observatory Photograph, courtesy of Richard Dreiser.

designated by their length rather than by their aperture!) This resulted in his first catalog of double and multiple stars. It also produced one of the greatest discoveries in the history of observational astronomy – made by a totally unknown "amateur!"

Uranus and The King's Astronomer

While sweeping the sky in the constellation of Gemini on the night of March 13, 1781, Herschel came across a small greenish disk of light. Careful observation showed that it was slowly moving among the stars, leading him to believe that it was a strange-looking comet. Others agreed with him and for nearly a year mathematicians attempted to calculate an orbit on that basis. All attempts failed and it was finally realized that Herschel had, in fact, found another planet! This was the first such world ever *discovered* (the five naked-eye planets Mercury, Venus, Mars, Jupiter, and Saturn having been known since antiquity) and it effectively doubled the size of the solar system. It apparently had never entered anyone's mind that there actually could be more planets lying beyond those already known.

This electrifying and unprecedented discovery catapulted Herschel to instant fame and brought him to the attention of King George III, who appointed him his private astronomer. This honor brought with it a salary sufficient to allow Herschel to give up his musical duties and spend full-time on astronomy. In gratitude, he named the new planet "Georgium Sidus" after his patron, but this did not find approval among other astronomers. "Uranus" was finally chosen instead, in keeping with the naming of the other five planets after gods of ancient mythology.

Caroline and Sir John

No account of the work of William Herschel is complete without mentioning his devoted sister, Caroline. She assisted him both at the telescope at night and in the arduous work of recording and reducing his many discoveries in preparation for their eventual publication. This, in addition to taking care of household duties including meals (even feeding and reading to her brother as he polished his mirrors for hours on end!). She became the leading woman astronomer of her day and the first to find a comet (her record of eight discoveries having stood for nearly two centuries). She observed with a small 27-inch focus Newtonian "comet sweeper" made expressly for her by William, using it to scan the sky on her own when he was away at meetings or showing the stars to the King and his court. Thus, a number of her own deep-sky discoveries are contained in Sir William's catalog – one of the author's personal favorites being the lovely rich open cluster designated HVI-30 (or NGC 7789) in Cassiopeia, to which I have given the name "Caroline's Cluster." (For more on this remarkable woman astronomer, see especially Michael Hoskin's *The Herschel Partnership: As Viewed by Caroline* listed in Appendix 2.) (Figs. 1.2, 1.3)

William Herschel married rather late in life, having a son named John Frederick William, or "John" for short. Like his father and his Aunt Caroline, he too became famous as an astronomer. But he was also a gifted mathematician and scientist in other fields, among other activities experimenting with photography and having taken the oldest existing photograph on a glass plate (a ghostly image of his father's 40-foot telescope!). John is best known for completing his father's survey of the northern sky and (especially) then extending it to the southern sky as well. He spent four years sweeping the heavens from Cape Town, South Africa, having taken his father's favorite telescope – the "Large" 20-foot reflector (see Chapter 2) – and logging thousands of previously unknown double stars, clusters, and nebulae. He returned to England in 1838, a national hero for this work, receiving among many other honors knighthood. Sir John eventually issued a catalog of his findings – and later a combined one containing all of his and his father's many telescopic discoveries. This latter work became the basis for the famed *New General Catalogue of Nebulae and Clusters of Stars* (or *NGC*), which was published in 1888 (Figs. 1.4, 1.5).

Fig. 1.2. Sadly, there are no pictures of Caroline Herschel as a young woman. However, this amazing reconstruction was done by an art student taking one of noted Harvard astronomer and science historian Owen Gingerich's astronomy courses. The artist also happened to have experience in making young students look old for theater plays, and said she simply reversed the process using published descriptions of the young Caroline. Painting by Lisa Rosowsky, 1987, courtesy of Dr. Owen Gingerich.

Explorer of the Heavens

Only a hint can be gleaned of the incredible life and works of this remarkable astronomer and his family from the necessarily brief account given here. You are strongly encouraged to consult the various Herschel references given in Appendix 2 – which make fascinating reading, especially on cloudy nights! But here, the author would like to share just two of the moving testimonials to be found in the biographies of William Herschel from this listing, in the hope of conveying at least some sense of this observer's rare genius and astounding accomplishments.

Lick Observatory Director Edward Holden, in his 1881 classic work *Sir William Herschel*, had this to say about the great astronomer and his observational model of the Milky Way:

> As a scientific conception it is perhaps the grandest that has ever entered into the human mind. As a practical astronomer he remains without an equal. In profound philosophy he has few superiors. By a kindly chance he can be claimed as the citizen of no one country. In very truth his is one of the few names which belong to the whole world.

Fig. 1.3. Caroline Herschel shown in her 80s back in Hanover, where she returned after Sir William's death. In a letter to her nephew, Sir John Herschel, she said: "You will see what a solitary and useless life I have led these 17 years all owing to not finding Hanover, nor anyone in it, like what I left, when the best of brothers took me with him to England in August, 1772." Courtesy of the Royal Astronomical Society/Science Photo Library, London.

Equally moving are the following words by British astronomical historian Agnes Clerke from her 1895 classic biography *The Herschels and Modern Astronomy*:

> The grand problem with which Herschel grappled all his life involves more compli-
> cated relations than he was aware of. It might be compared to a fortress, the citadel
> of which can only be approached after innumerable outworks have been stormed.
> That one man, urged on by the exulted curiosity inspired by the contemplation of the
> heavens, attempted to carry it by a *coup de main*, and, having made no inconse-
> quential breach in its fortifications, withdrew from the assault, his 'banner torn, but
> flying,' must always be remembered with amazement.

Fig. 1.4. Sir John Herschel seen in his prime. He not only extended his father's work in the northern skies over England, but also explored those of the southern heavens as well from Cape Town, South Africa, using Sir William's favorite telescope – the "large 20-foot" reflector (see Fig. 2.2). Courtesy of the Royal Astronomical Society/ Science Photo Library, London.

Fig. 1.5. John Herschel in his later years. In addition to his fame as an astronomer, he was also a brilliant mathematician, gifted science writer, and a pioneer in the field of photography. Courtesy of the Royal Astronomical Society/Science Photo Library, London.

Chapter 2

Herschel's Telescopes

Early Instruments

William Herschel began his telescope-making career in 1773 by experimenting with relatively small refractors (small in aperture, but certainly not in length – one of them being 30 feet long!). As these were then still optically primitive compared to today's instruments, he soon turned his attention toward reflectors. These could be made in larger sizes and without concern for the quality of optical glass, for they used mirrors instead of lenses. But these were not the familiar telescope mirrors of today, as silver-on-glass optics did not appear until long after Herschel's death. Instead, they were made of speculum metal – a brittle and hard casting composed mainly of copper and tin. He first made several mirrors for a 5.5-foot Gregorian reflector, but then turned to the simpler Newtonian form. All of his telescopes from that point on were long-focus Newtonians of ever-increasing size, culminating in the great 40-foot reflector (see below).

He soon produced a 7-foot telescope (as previously mentioned, telescopes in Herschel's day were specified by their length rather than by the size of their optics), probably of around 6-in. aperture. He also made several 9-inch mirrors for a 10-foot reflector (and much later a 10-foot reflector with a 24-inch mirror), followed by 20-foot models having 12- and 18.7-inch mirrors as described below. But his favorite early reflector was another 7-foot which contained "a most capital speculum" as he described it of 6.2-inch aperture. This is the telescope that he used for his first "review" of the heavens and the one with which he discovered the planet Uranus (Fig. 2.1).

A Telescope-Making Business

Herschel's telescopes far surpassed in both quality and size any other telescope in the world at that time. After comparison trials at a number of observatories in England including Greenwich, he stated with confidence "I can now say that I absolutely have the best telescopes that were ever made." His fame as a telescope-maker spread rapidly and soon he was flooded with requests from both other observers and observatories to make instruments for them. While he was not in the telescope-making business as such, his allowance from the King – while freeing him from his musical duties – did not entirely meet his expenses, and so he began to make and sell telescopes privately. In addition to at least 60 complete instruments (most of them 7- and 10-feet in size), he also made several hundred mirrors upon order in addition to those for his own telescopes!

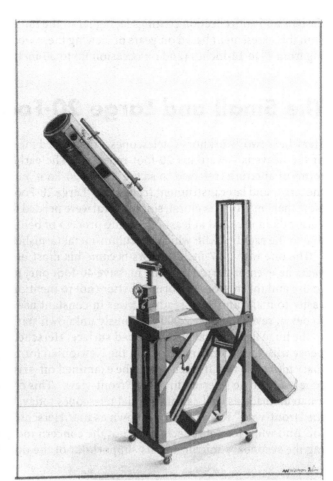

Fig. 2.1. The 6.2-in. "7-foot" reflector with which William Herschel discovered the planet Uranus on the night of March 13, 1781. Like all of his telescopes, it employed a speculum-metal mirror that was ground, polished, and figured with his own hands. Courtesy of the Royal Astronomical Society/Science Photo Library, London.

While most of the objects Herschel discovered in his first review of the heavens with his prized 7-footer were double and multiple stars, he also found a number of his early clusters and nebulae with it. (It should be mentioned here that in his day and long thereafter, galaxies were not yet recognized as such, being simply lumped under the category of "nebulae.") It is frequently stated that a modern 6- to 8-inch telescope will show a large percentage of the objects in the Herschel catalog (including many of his faint and very faint nebulae), and that a good 12-inch telescope should reveal every one of them even though most were found using his two 20-foot instruments. This is largely possible, of course, due to the much higher reflectivity of today's coated-glass telescope mirrors – and to a lesser extent to modern eyepieces as well. (Herschel primarily used single-lens oculars,* multiple-element

---

*As a fascinating aside, Herschel often mentioned using very high magnifications for his solar system and double star studies employing these simple eyepieces – in some cases in excess of 6,000×! While many in his time doubted these claims, modern optical tests on his eyepieces prove that he did, indeed, achieve such remarkable powers. One of his surviving oculars actually has a focal length of just 0.0111 in.! But no one was more aware of the limitations of high powers than he, and most of his observing (even with his large instruments) was conducted at magnifications under 300×.

designs and antireflection coatings laying far in the future.) The author fully agrees with this assessment based on years of viewing these wonders with telescopes ranging from 2- to 14-inches (and on occasion up to 30-inches!) in aperture.

The Small and Large 20-Foot

Herschel's two "workhorse" telescopes – those used for all his later various reviews of the heavens – were his 20-foot reflectors. The earlier and smaller of these in terms of aperture (referred to as the "Small 20-Foot") used 12-inch mirrors, while the larger and later instrument (called the "Large 20-Foot") used 18.7-inch mirrors. Note that "mirrors" is plural, since several were needed for each telescope – the one currently in use, and at least one in the process of being repolished and refigured due to the rapidity with which speculum-metal tarnished (Fig 2.2)!

The one with 18.7-inch mirrors became his most useful telescope and in later years he even preferred it to the massive 40-foot one, for it was both much easier to use and the mirrors performed better (not to mention that they were also vastly easier to make and keep ready). It was in constant use on clear nights from dusk to dawn, revealing over 2,000 previously unknown star clusters and nebulae. Due to the huge light-loss at each reflected surface, Herschel eventually decided to dispense with the secondary mirror in the Newtonian form. Instead, he tilted the primary mirror so that its focus could be examined off-axis directly at the front of the tube – a form he referred to as the "front-view." This concept is still used in some amateur-made as well as commercial telescopes today, but instead of being called the "front-view" form it is now known as the "Herschelian" in honor of its inventor. And while loss of reflectivity is not the concern today it was in his time, moving the secondary mirror and its support out of the optical path essentially gives

Fig. 2.2. The "Large 20-foot" reflector had an aperture of 18.7 inches This was Sir William's most useful instrument and the one with which he discovered most of his clusters, nebulae, and galaxies. (He called it "large" to distinguish it from another earlier 20-foot reflector that used a 12-inch mirror.) Courtesy of the Royal Astronomical Society/Science Photo Library, London.

the unobstructed performance of a refractor combined with total freedom from the color aberrations inherent in lenses.

The Great 40-Foot

Herschel's most ambitious telescope-making project – indeed, the most ambitious in history up to that time – was the construction of his great "40-Foot" reflector with its 48-inch (or 4-foot) diameter mirror (resulting in a focal ratio of $f/10$). He received financial support for this massive undertaking from the King, as well as an annual allowance for upkeeping the telescope once it was completed. Herschel actually made several mirrors for it before he finally was able to get one that would take an acceptable polish and figure (Fig 2.3). (Interestingly, Herschel had much earlier tried to make a mirror for a proposed 30-ft telescope, but gave up on the idea after several near-disastrous events that occurred during attempts at casting it.)

In 1787, Herschel climbed into the mouth of the huge tube and searched for the focus using one of his first mirrors. His target was the Orion Nebula, which he described as "extremely bright" but the figure was far from perfect. On later attempts he used Saturn as his test object, discovering several new satellites while at it. Some idea of the light-grasp of this instrument can be had from this famous description by Sir William of the star Sirius as seen through it:

> ... the appearance of Sirius announced itself, ... and came on by degrees, increasing in brightness, till this brilliant star at last entered the field of view of the telescope, with all the splendour of the rising sun, and forced me to take the eye from that beautiful sight.

Regular work with the telescope finally began in 1789. But Herschel was never pleased with the telescope's performance. Perhaps this is best summed up in the following lines from telescope historian Henry King's definitive work, *The History of the Telescope*:

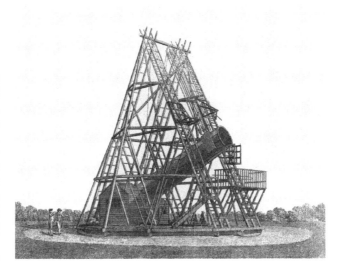

Fig. 2.3. Herschel's great 40-foot reflector which housed a 48-inch diameter mirror. This wonder of the ages attracted royalty, dignitaries, and other visitors from far and wide. Even today, this famous image remains a lasting icon to a bygone era of visual observational astronomy. Courtesy of the Royal Astronomical Society/ Science Photo Library, London.

The paucity and irregularity of Herschel's observations with the 40-foot leave little doubt that the great telescope failed to meet its maker's expectations. In the first place, the weather was seldom good enough to allow full use of its aperture and, when conditions were favorable, Herschel preferred the smaller and more manageable 20-foot. He found there were few objects visible in the 40-foot which he could not see in its smaller counterpart.

That very few of the objects contained in the Herschel catalog were actually discovered with the 40-foot certainly confirms the above statement. How very sad for Herschel after all his labors over this great instrument! But while it was a disappointment for him, it was certainly not for the many sightseers who came to gawk at this wonder of the ages, including royalty and dignitaries of all levels and noted scientists from the world over. One famous early event from this period involves the day the King came to inspect the telescope while still under construction, bringing with him the Archbishop of Canterbury. As they were about to enter the open mouth of the tube (which at this point still lay on the ground), the King said "Come, my Lord Bishop, I will show you the way to Heaven!" Even today, the image of Herschel's mammoth 40-foot telescope remains one of the great – if not *the* greatest – icons of astronomical history.

In closing, two very important points need mention. First, all of Herschel's many telescopes were mounted as simple altazimuths, being moved about the sky and tracked manually. And secondly, they were all mounted outside of his various residences in the open night air. For all his fame and discoveries, Sir William never had an observatory!

Chapter 3

Herschel's Catalogs and Classes

Double and Multiple Stars

Everywhere in the sky where William Herschel pointed his mighty telescopes there were wonders no one before him had ever seen and he truly had a boundless untrodden field before him. He left the world with two great catalogs of his astronomical discoveries resulting from these heady celestial explorations. The earlier of these contain the double and multiple stars found during his sweeps of the heavens, the first of which was conducted with his favorite 6.2-inch 7-foot reflector at a fixed magnification of 227×. (As previously mentioned, this is the same instrument that was used for his discovery of the planet Uranus.) During later reviews with both the 7-foot and his two 20-foot telescopes, he found many more – his total number of discoveries being in excess of 800 pairs. (One of his most spectacular finds is the beautiful triple system β Monocerotis, better-known as Herschel's Wonder Star.) He would often examine more than 400 stars a night, looking for duplicity, multiplicity, and anything else noteworthy such as marked colors.

Herschel grouped his many double star discoveries into the following six classes (somewhat reminiscent of his classes for clusters and nebulae, discussed below) in the several catalogs he issued between the years 1782 and 1784:

H I – difficult to resolve and/or measure
H II – close but measurable
H III – 5″ to 15″ separation
H IV – 15″ to 30″ separation
H V – 30″ to 60″ (1′) separation
H VI – 1′ to 2′ separation

Note that Herschel used the prefix "H" to distinguish his discoveries from those of his son John (designated by "h") – which for the northern and southern sky combined, number well into the thousands! (Just one outstanding example of the latter's findings is the magnificently-hued orange and blue pair h3945, located southeast of Sirius in Canis Major – an object I have dubbed the Winter Albireo from its resemblance to that famous namesake in Cygnus.) An additional catalog of new doubles found by Sir William appeared much later, in 1821, using the prefix "H N" to distinguish them from his earlier discoveries.

Incidentally, readers interested in observing these tinted jewels of the sky are referred to the author's book *Double and Multiple Stars and How to Observe Them* (Springer, 2005).

Star Clusters, Nebulae and Galaxies

Despite Herschel's pioneering discoveries in the field of stellar astronomy, it is his deep-space explorations for which he is best-known and remembered. The more than 2,500 star clusters and nebulae (which included many galaxies, the true nature of which was unrecognized at that time) were cataloged under the following eight categories or "Classes" as he called them, with the total number of objects in each indicated in parentheses:

Class I – Bright Nebulae (288)
Class II – Faint Nebulae (909)
Class III – Very Faint Nebulae (984)
Class IV – Planetary Nebulae (78)
Class V – Very Large Nebulae (52)
Class VI – Very Compressed and Rich Clusters of Stars (42)
Class VII – Compressed Clusters of Small (Faint) and Large (Bright) Stars (67)
Class VIII – Coarsely Scattered Clusters of Stars (88)

Thus, Herschel's entire catalog contains a total of 2,508 *entries*, with Classes II and III accounting for 1,893 of them. (Note that the actual *number* of objects is somewhat less than this, since some three dozen were either inexplicably assigned to more than one Class or were entered twice in the same Class by Sir William.) Such a large number of targets (with a great percentage of them being labeled faint and very faint by their discoverer who used the largest telescopes in the world at the time to find them) have discouraged most observers from attempting to view the entire catalog. As mentioned previously in the Preface, the author had suggested some years ago in articles and letters in both *Sky & Telescope* and *Astronomy* magazines that a much more realistic goal could be had by dropping Classes II and III as largely difficult and visually less-interesting specimens and going after the remaining 615 objects. This suggestion was the motivation for the founding of a national Herschel Club, as discussed in both the Preface and also in Appendix 1. Chapters 5 through 10 highlight a total of some 165 showpieces contained in the six remaining Classes. These not only provide a great sampling of Herschel objects in themselves, but they are also ideal as preparation for those observers who are considering viewing all 615 targets (a complete listing of which will be found in Appendix 3). (Fig. 3.1).

Out of respect for Messier's work, Herschel included very few of the famed M-objects in his own compilations, and those that he did were mostly ones numbered from M104 to M110. These objects were only attributed to Messier long after Herschel's time, later historical research showing that they had indeed been seen by Messier (or one of his colleagues) but were not included in his original catalog. One rather surprising exception to this rule is H V-17, which is actually

Fig. 3.1. Three famous deep-sky observing guides from the past: left, *The Bedford Catalog* from *A Cycle of Celestial Objects* by W.H. Smyth; middle, Volume Two of *Celestial Objects for Common Telescopes* by T.W. Webb; right, *1001 Celestial Wonders* by C.E. Barns. All of these classic works list objects by their Herschel class and number. Photo by Sharon Mullaney.

the big spiral galaxy M33 in Triangulum. There has been some speculation that what Herschel was really recording was not M33 itself but rather one of the glowing star-clouds in its spiral arms designated NGC 604. However, he actually entered this object separately as H III-150, so the mystery of why M33 was included in his list remains. Two other exceptions having Herschel designations are the peculiar edge-on galaxy M82 and the Trifid Nebula (M20).

Incidentally, mentioning Messier's list brings to mind the much more recent Caldwell Catalog compiled by Sir Patrick Moore of England. Like the earlier listing (allowing for the duplication of M102), it contains 109 deep-sky targets. Although a number of Herschel objects are included among them, the Caldwell roster offers only a sampling of the many unsung wonders awaiting observers in Sir William's catalog. And it also, unfortunately, passes over some really great showpieces – just one example being the bright spiral galaxy H I-56/57 (NGC 2903/5), located just off the "Sickle" of Leo and described in Chapter 5 (Fig. 3.2).

It should be noted that no Herschel object (at least not those of William!) will be found below a Declination of about −33° (actually −32° 49′ – his southernmost discovery being the galaxy H I-241/NGC 3621 in Hydra) due to the rather high latitude of his various observing sites around London. Had he been able to see another 7°–10° or so further south, he would surely have laid claim to (and been thrilled by!) such wonders as the big, bright galaxies NGC 1316 and NGC 1365 in Fornax, and NGC 55 and NGC 300 in Sculptor; or the radiantly glorious open cluster NGC 2477 in Puppis; or the fascinating planetaries NGC 6302 in Scorpius (the Bug Nebula) and NGC 3132 in Vela (the Eight-Burst Planetary – rival of the northern sky's famed Ring Nebula in Lyra). Instead, these were left to his son to find during his later survey of the southern sky.

Fig. 3.2. The massive three-volume *Burnham's Celestial Handbook* by Robert Burnham, Jr. contains over 2,100 pages and covers some 7,000 celestial wonders! Like the classics shown in Fig. 3.1, it also lists star clusters, nebulae, and galaxies by their Herschel classes and numbers – and in this case, by their NGC and IC designations as well. The copies shown are well-worn from constant use by the present author! Photo by Sharon Mullaney.

Miscataloged Objects

A fair number of William Herschel's discoveries were cataloged in the wrong Class. As just one striking example, there is H IV-50 (NGC 6229) in Hercules, which is described in Chapter 6. He considered this tiny object to be a planetary nebula and so it was long viewed to be such by many of the classic observers. And indeed, it sure does look like a typical planetary in the eyepiece. But in actuality, it is a globular cluster that apparently was beyond the resolution capabilities of his telescopes. A look toward the end of the list of "bright nebulae" in Class I in Appendix 3 will reveal many other small globular clusters assigned there as nebulae. Likewise, the listing given under "planetary nebulae" in Class IV actually contains many diffuse nebulae and more galaxies than true planetaries. Such misidentifications as these are hardly to be unexpected, for nothing was known at the time of the actual physical nature of most of the nebulous objects he found – spectroscopic analysis and astrophysics still lying in the future. Herschel based the classification of his discoveries *solely upon their visual appearance in his various telescopes.* But all of this adds to the fascination of viewing these objects, as observers attempt to verify Sir William's eyepiece impressions and to see why he placed given objects in the classes he did.* And while having to rely strictly on

---

*It is very important here to note that in compiling the *NGC*, Dreyer frequently added to and/or modified Herschel's original shorthand descriptions in light of more recent knowledge. This helps to explain at least some of the enigmas encountered – as, for example, when an object is stated to be a globular cluster but Sir William does not mention seeing stars and assigns it to one of his classes of nebulae. *This should be kept in mind whenever reading any of the descriptions of his various discoveries contained in Part II of this book.*

what they looked like to the eye resulted in a sizeable number of objects being misclassified, in other cases this reliance led to some amazing insights and discoveries! One particularly striking example involves the planetary nebula H IV-69 (NGC 1514) in Taurus – a story reserved for Chapter 6.

The GC, NGC, and NGC 2000.0

In closing, mention should be made that William Herschel's various lists of double star and deep-space discoveries originally appeared as papers in the *Philosophical Transactions* of the Royal Society of London in the late 1700s. In 1864, John Herschel published his monumental *General Catalogue of Nebulae* (or *GC*) also in the *Transactions*, its more than 5,000 nonstellar objects having been found mostly by his father and himself. Based heavily on the *GC*, J.L.E. Dreyer subsequently compiled the famed *New General Catalogue of Nebulae and Clusters of Stars* (or *NGC*). Its more than 7,800 entries contain among them the only complete listing of all of Sir William's discoveries outside of the *GC* itself, giving his numbers by Class for each object and a shorthand descriptive notation that he and his son invented (and which was adopted by Dreyer for the entire *NGC*) (Fig. 3.3).

In 1953, and again in 1962 and 1971, the Royal Astronomical Society in London reprinted as a single volume the *NGC* combined with its two later additions known as the *Index Catalogues*. Sadly, this treasured work has once again gone out-of-print

Fig. 3.3. The old and the new! On the left is J.L.E. Dreyer's original *NGC* shown in its reprinted edition by the Royal Astronomical Society (London), while on the right is Roger Sinnott's *NGC 2000.0*. The latter work contains updated coordinates, object types, magnitudes and angular sizes in addition to the *NGC*'s original shorthand descriptions. Unfortunately, the Herschel classes and numbers have been replaced by the more modern NGC numbers. Photo by Sharon Mullaney.

and is now only to be found in observatory libraries and on the rare/used-book market. In 1988, Roger Sinnott, a longtime editor at *Sky & Telescope* magazine, performed a great service to the astronomical community by issuing his *NGC 2000.0* (Cambridge University Press and Sky Publishing). While bringing this seminal work up to date with Epoch 2000.0 coordinates, modern object-types, angular sizes and magnitudes, as well as correcting many errors, Sinnott unfortunately dropped the columns giving the names and catalog numbers of the original discoverers – including those of both the Herschel's. Thus, the *NGC* itself still remains the ultimate reference for we Herschelians!

Chapter 4

Observing Techniques

Is This Really Necessary?

Two famous statements made by William Herschel underscore the importance of this lengthy chapter as preparation for observing the Herschel objects. One is that "Seeing is in some respect an art, which must be learnt." As described below, there is no doubt that the human eye can be trained to see better in at least four distinct areas involving the viewing of celestial wonders – these being dark adaptation, averted vision, color perception, and visual acuity. And the reason that this is really possible is that the eye works not alone but in conjunction with an amazing "image-processing computer" – the human brain!

Herschel's other dictum is that "When an object is once discovered by a superior power [a large telescope], an inferior one [a smaller telescope] will suffice to see it afterwards." The truth of this statement has been demonstrated countless times by visual observers from Sir William's time down to the present. One classic case involves the famed white-dwarf companion to the star Sirius. It took an 18-inch refractor to discover this minute object, but (depending on just where it is in its tight orbit about the primary) it can and has been seen in instruments as small as 4- to 6-inches in aperture!

The time spent studying and applying what follows will pay off significantly by enhancing your exploration and enjoyment of the wonders contained in the Herschel catalog. Readers desiring to delve even deeper into such matters – including the fundamentals of telescopes, eyepieces, mountings, and accessories, as well as their use – should consult the author's book *A Buyer's and User's Guide to Astronomical Telescopes and Binoculars* (Springer, 2007).

Dark Adaptation

It is an obvious fact that the eyes need time to adjust to the dark after coming out of a brightly-lit room – a phenomenon known as *dark adaptation*. And two factors are actually at play here. One is the dilation or opening of the pupils themselves, which begins immediately upon entering the dark and continues for several minutes. The other involves the actual chemistry of the eye, as the hormone rhodopsin (often called "visual purple") stimulates the sensitivity of the rods to low levels of illumination. The combined result is that night vision improves noticeably for perhaps half an hour or so (and then continues to do so very slowly for many hours following this initial period). This is why the sky typically looks black on first

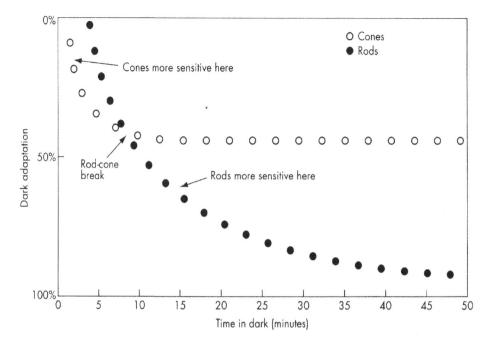

Fig. 4.1. Dark adaptation times for both the color-sensitive cones (open circles) at the center of the eye, and the light-sensitive rods (black dots) around the outer part of the eye. At the "rod-cone break" some 10 min after being in the dark, the sensitivity of the cones levels off and remains unchanged. The rods, however, continue to increase their sensitivity to low light levels, with complete dark adaptation taking at least 4 hours! For all practical purposes, the eye is essentially dark-adapted in about 30–40 minutes. Good dark adaptation is essential for viewing the fainter Herschel wonders.

going outside, but later appears gray as you fully adjust to the dark. In the first instance, it is a stark contrast effect and in the second the eye has become sensitive to surrounding illumination, light pollution, and the natural airglow of the night sky itself that were not seen initially (Fig. 4.1).

Stargazers typically begin their observing sessions by viewing bright objects like the Moon and planets first and moving to fainter ones afterwards, giving the eye time to gradually dark-adapt naturally. This procedure is of critical importance in observing deep-sky objects like those in the Herschel catalog. Stars themselves are generally bright enough that they can be seen to full advantage almost immediately upon looking into the telescope. (Exceptions are faint pairs of stars and dim companions to brighter stars, where the radiance of the primary often destroys the effect of dark adaptation.) White light causes the eye to lose its dark adaptation but red light preserves it, making it standard practice to use red illumination to read star charts and write notes at the eyepiece. Another helpful technique is to wear sunglasses (preferably polarized) whenever venturing outside on a sunny day if you plan to look for "faint fuzzies" that evening. It has been shown that bright sunlight – especially that reflected from an ocean beach, bodies of water, and from snow – can retard the eye's dark adaptation for as long as several days!

Herschel often placed a dark hood over his head to maintain his dark adaptation. This is also done by some observers today, especially those living in heavily light-polluted areas. Such "photographer's cloths" (as they are generally known) are simply dark opaque pieces of fabric that are thrown over the observer's head and the eyepiece area of the telescope, effectively eliminating stay light and preserving dark adaptation. They are available from camera stores and some telescope dealers, and can also be easily fabricated. In actual practice, these hoods can prove a bit suffocating – especially on warm muggy nights – and are sure to raise the eyebrows of any neighbor who happens to see you lurking in the dark!

Averted Vision

A second area of training the eye-brain combination involves the technique of using *averted* (or side) *vision* in viewing faint celestial objects. This makes use of the well-known fact that the outer portion of the retina of the eye contains receptors called *rods*, and that these are much more sensitive to low levels of illumination that is the center of the eye whose receptors are known as *cones*. (See the discussion below involving color perception by the cones.) This explains the common experience of walking or driving at night and seeing objects out of the corner of your eye appearing brighter than they actually are if you turn and look at them directly.

Applied to astronomical observing, averted vision is used in detecting faint companions to double stars and dim stars in open and globular clusters. But it is especially useful (and its effect most obvious) in viewing low-surface-brightness targets like the nebulae and galaxies to be found in the Herschel catalog. Here, increases in apparent brightness from two to two- and a-half times (or an entire magnitude), have been reported! Once having centered such an object in the field of view, look to one side of it (above or below also works), and you will see it magically increases in visibility. (Be aware that there is a small dark void or "blind spot" in the retina between the eye and ear that you may encounter in going that direction.) (Fig. 4.2)

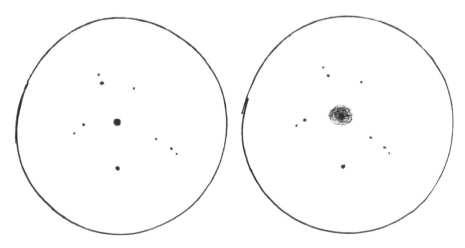

Fig. 4.2. Staring directly at the Blinking Planetary (H IV-73/NGC 6826) reveals only its bright central sun, as shown at left. Upon switching to averted vision, the nebulosity suddenly appears and all but drowns out the star itself, as shown at right!

One of the most dramatic examples of the effect of averted vision involves the bright planetary nebula H IV-73 (NGC 6826) in Cygnus. Better-known as the "Blinking Planetary" (a name coined by the author many years ago in *Sky & Telescope* magazine), this amazing object is described in some detail in Chapter 6. It is truly one of the most exciting sights in the entire Herschel catalog, for it is one of the few deep-sky wonders that appear to "do something" as you watch it!

Color Perception

A third area involving the eye-brain combination is that of *color perception*. At first glance, all stars look white to the eye. But upon closer inspection, differences in tint among the brighter ones reveal themselves. The lovely contrasting hues of ruddy-orange Betelgeuse and blue–white Rigel in the constellation Orion is one striking example in the winter sky. Another can be found in the spring and sum-mer skies by comparing blue–white Vega in Lyra, orange Arcturus in Bootes, and ruddy Antares in Scorpius. Indeed, the sky is alive with color once the eye has been trained to see it! (Star color, by the way, is primarily a visual indication of surface temperature: reddish stars are relatively cool while bluish ones are quite hot. Yellow and orange suns fall in between these extremes.)

While color perception many not seem to be a concern in viewing Herschel objects, it actually is for perceiving the tints of stars in open clusters and the eerie hues of many of the planetary nebulae he discovered (which often shine with vivid combina-tions of blues and greens). While the rods on the edge of the eye are light sensitive, they are essentially colorblind. Thus, for viewing the tints of these and other celestial wonders, direct vision is employed – making use of the color-sensitive cones at the center of the eye. The rule is simple: stare directly at an object to perceive its color and off to the side to see it become brighter (unless it is already a bright target like a planet or naked-eye star). Incidentally, mention should be made here of a very peculiar phe-nomenon known as the "Purkinje Effect" which results from staring at red stars – they appear to increase in brightness the longer you look at them!

Visual Acuity

The fourth area of training the eye–brain combination is that of *visual acuity* – the ability to see or resolve fine detail in an image. There is no question that the more time you spend at the eyepiece, the more such detail you will eventually see! Even without any purposeful training plan in mind, the eye–brain combination will learn to search for and see ever-finer detail in what it is viewing. But this process can be considerably speeded up by a simple exercise repeated daily for a period of at least several weeks. On a piece of white paper, make a circle about 3 inches in diameter. Then using a soft pencil randomly draw various markings within the circle, ranging from broad patchy shadings to fine lines and points. Now place the paper at the opposite end of a room at a distance of at least 20 feet or so, and begin sketching what you see using the unaided eye. Initially, only the more prominent markings will be visible to you – but as you repeat this process over a period of time, you will actually see more and more of them!

Tests have shown improvements in overall visual acuity by a factor of as much as 10 as the eye is trained! Not only will you see more detail on the Sun, Moon, and planets as a result of this practice, but you will also be able to split much closer double stars than you were previously able to. As visual acuity relates to observing the Herschel objects, you will be able to better resolve the individual stars in small, compact open clusters and tight globulars. It will also help you to see intricate details in the brighter diffuse and planetary nebulae, and to pick out features like the nucleus, spiral arms, and star-forming regions in galaxies.

Magnification and Field of View

It is generally recognized that the higher the magnification used on a telescope, the smaller the actual amount of the sky (or field of view) that is seen. For this reason, it is standard procedure in sweeping for deep-sky wonders like those in the Herschel catalog (especially large star clusters, extended nebulosities, and big galaxies) that the lowest possible power giving the widest possible view be used. Once found, higher magnifications can then be employed for taking a more detailed look at the object if desired. This is especially useful in the case of tight globular clusters, small planetary nebulae, and for seeing structure in galaxies. Recommended scanning powers for a typical telescope are 7× to 10× per inch of aperture – or 35× to 50× in the case of a 5-inch glass. Medium magnifications generally run between 15× and 25× per inch, while the practical upper limit most nights is 50× per inch of aperture. (The magnification of a telescope is found by dividing its focal length by the focal length of the eyepiece used, both units being either in millimeters or inches.) One of the author's favorite telescopes (mainly due to its extreme portability) happens to be a 5-inch f/10 Schmidt-Cassegrain catadioptric on a lightweight altazimuth mounting. For sweeping, 40× is used and for closer looks 80× to 100×. The resulting field of view ranges from over 1° at low power to around 30′ at medium ones, depending on the actual eyepiece type used. Higher magnifications are reserved mainly for the Moon, planets, and close double stars on steady nights (Fig. 4.3).

Fig. 4.3. As a telescope's magnification increases, the actual amount of the sky seen decreases (making low powers preferred for many types of observing, such as sweeping for the Herschel objects). Shown here are three stylized sketches of the Moon at low, medium, and high magnifications. While the image gets bigger at higher powers, less and less of it can be fitted within the eyepiece's field of view.

Those *actual fields* of view quoted are based on using eyepieces having 50° *apparent fields* (which is the angle you see looking through the eyepiece itself at the daytime sky).

Dividing the magnification into the eypiece's apparent field gives the actual field. Thus, an ocular having an apparent field of 50° that produces 50× on a given telescope results in an actual field of 1° – or two full-Moon diameters of sky. But as most readers are no doubt aware, many wonderful new eyepiece designs are widely available today (some at considerable cost!) having apparent fields from 60° all the way up to a whopping 82°! These produce marvelous "spacewalk" views of the heavens that must be seen to be believed. As such, these wide-angle and super-wide-angle oculars are much to be desired for sweeping the sky. We can only wonder what Sir Willliam would have thought of these optical marvels, for he mainly used single-element eyepieces that gave actual fields of only a fraction of a degree on his various telescopes!

Incidentally, some observers prefer to sweep with more magnification than that recommended above, for higher powers darken the sky background. But again, these higher powers bring with them reduced fields of view. At the other extreme are short-focus, rich-field telescopes (or RFTs) producing actual fields of 2° to 3° in extent. They are quite marvelous for general sweeping of the sky, and for viewing big star clusters like the Pleiades, Hyades, and Beehive, or for sweeping the Milky Way's starclouds. However, their inherently small image scales make it easy to pass right over tiny objects like typical planetary nebulae or dim galaxies.

Sky Conditions

A number of atmospheric and related factors affect the visibility of celestial objects at the telescope. In the case of the Moon, planets, and double stars the most important of these is atmospheric turbulence or *seeing*, which is an indication of the steadiness of the image. On some nights, the air is so unsteady (or "boiling" as it is sometimes referred to) that star images appear as big puffy, shimmering balls, and detail on the Moon and planets is all but nonexistent. This typically happens on nights of high atmospheric *transparency* – those having crystal-clear skies, with the air overhead in a state of rapid motion and agitation. On other nights, star images are nearly pinpoints with virtually no motion, and fine detail stands out on the Moon and planets like an artist's etching. Such nights are often hazy and/or muggy, indicating that stagnant tranquil air lies over the observer's head and that seeing is superb (Fig. 4.4).

Various seeing and transparency scales are employed by observers to quantify the state of the atmosphere. One of the most common of these uses a 1-to-5 numerical scale, with 1 indicating hopelessly turbulent blurred images and 5 stationary razor-sharp ones in the case of seeing. The number 3 denotes average conditions. Others prefer a 1-to-10 system, with 1 again representing very poor and 10 virtually perfect seeing, respectively. Transparency is usually rated on a 1-to-5 scale, with 1 representing heavy haze/humidity and 5 superbly clear skies. (In some schemes, the numerical sequence is reversed, with lower numbers indicating better and higher numbers, poorer conditions.) While casual stargazing can often be done on nights of less than average-quality image-steadiness, observations requiring optimum resolution need good to excellent conditions. And for viewing

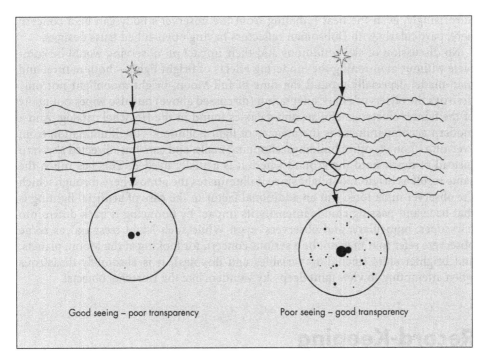

Fig. 4.4. On nights of good "seeing" or atmospheric steadiness, the air above the observer is tranquil and lies in relatively smooth layers, often resulting in somewhat hazy skies. This allows starlight to pass through undisturbed and produces sharp images at the eyepiece. In poor seeing the atmosphere is very turbulent, typically bringing with it crystal-clear skies and an ideal time to view faint objects. But the resulting star images are often blurry shimmering balls of light, making such nights largely useless for seeing fine lunar and planetary detail, for splitting close double stars, and for resolving tight globular clusters.

"faint fuzzies," good transparency is an absolute "must." On the other hand, Sir William worked on *any* night when the sky was clear *whatever* its condition!

Another factor affecting telescopic image quality is that known as "local seeing" – or the thermal conditions around and within the telescope itself. Heat radiating from driveways, walks and streets, houses, and other structures (especially on nights following hot days), plays a significant role in destroying image quality that is totally unrelated to the state of the atmosphere itself. For this reason observing from fields or grassy areas away from buildings and highways gives the best chance of minimizing local seeing caused by such sources.

There is also the matter of the cooling of the telescope optics and tube assembly themselves, which is especially critical for achieving sharp images. Depending on the season of the year, it may take up to an hour or more for the optics (especially the primary mirror in larger reflectors) to reach equilibrium with the dropping night air temperature. During this cool-down process, air currents within the telescope tube itself can play absolute havoc with image quality, no matter how good the atmospheric seeing is – even in closed-tube systems like the popular Schmidt-Cassegrain catadioptric. Reflecting telescopes should have tubes at least several inches larger than the primary mirror itself to allow room for thermal currents to rise along the inside of the tube rather than across the light path itself. (Herschel's 40-foot used a 5-foot diameter tube to house its 4-foot mirror.)

Surprisingly, even the heat radiating from the observer's body can be a concern here, particularly with Dobsonian reflectors having open-tubed truss designs.

No discussion of sky conditions and their impact on observing would be complete without mentioning the hindering effects of bright lights – both natural and man-made. Especially around the time of full-Moon, bright moonlight not only destroys the observer's dark adaptation (discussed above) but also wipes out many of the fainter clusters, nebulae, and galaxies found in the Herschel catalog. And a modern accompaniment is the menace of light pollution – the illumination from ever-more homes, office buildings, shopping malls, and cars lots, directed skyward instead of down onto the ground where it is actually needed. This has much the same result as bright moonlight since it illuminates the atmosphere through which the observer must look. But an additional factor in the case of artificial lighting is that haze and passing clouds intensify its impact by bouncing it back down into telescopes, binoculars, and observers' eyes. While such "light trespass" as some observers refer to it may not be a serious concern for looking at the Moon, planets, and brighter stars (including variables and doubles), it is absolutely disastrous when attempting to view faint deep-sky wonders like the Herschel objects!

Record-Keeping

The annuals of both amateur and professional astronomy attest to the personal as well as the scientific value of keeping records of our nightly vigils beneath the stars. William Herschel himself was scrupulous in this regard and employed a full-time assistant (his devoted sister Caroline!) just to record his nightly observations as he was making them at the telescope. From the former perspective, an account of what has been seen each night can bring pleasant memories as we look back over the years at our first views of this or that celestial wonder – or when we shared their very first look at the Moon or Jupiter or Saturn with loved ones, friends, and even total strangers. Our eyepiece impressions written and/or sketched on paper, or perhaps recorded on audio tape and/or electronically imaged, can provide many hours of nostalgic enjoyment in months and years to come. As far as observing the Herschel objects is specifically concerned, you are required to keep a log of your observations of each object on the Herschel Club target list in order to qualify for the coveted Herschel Certificate (see Appendix 1). (Fig. 4.5)

From a scientific perspective, often has the call gone out to the astronomical community in the various magazines, journals and electronic media asking if anyone happened to be looking at a certain object or part of the sky on a given date and at a particular time. If you happened to be at "the right place at the right time" indicated but noted nothing unusual in your observing log, that is still a fact of real importance to researchers. (This frequently happens in attempting to pin down when a nova in our galaxy or a supernova in a neighboring galaxy first erupted.) And, of course, there is always the possibility that you will be the first to see and report something new in the sky yourself!

The information in your logbook should include the following: the date, and beginning and end times of your observing session (preferably given in Universal Time/Date); telescope size, type and make used; magnification/s employed; sky conditions (seeing and transparency on a 1-to-5 or 1-to-10 scale, along with notes on passing clouds, haze and moonlight, or other sources of light-pollution); and

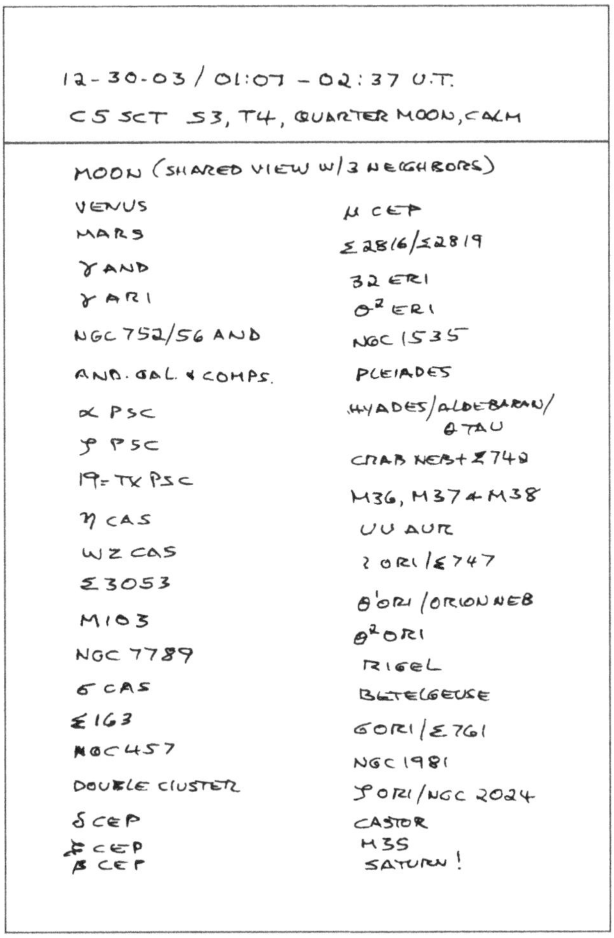

Fig. 4.5. A busy night's entry from the author's personal observing logbook. The date and times are given in Universal Time (U.T.) – that of the Greenwich, England, time zone. A 5-inch Celestron Schmidt-Cassegrain catadioptric telescope (C5 SCT) was used under conditions of average seeing (S) and good transparency (T), and the sky was brightened by the light of a first-quarter Moon. All of the targets viewed on this particular night are celestial showpieces! Normally, more time would be given to viewing fewer objects than shown here in order to fully enjoy and appreciate the cosmic pageantry.

finally a brief description of each object seen. And here, an important point should be borne in mind regarding record-keeping at the telescope: *limit the amount of time you spend logging your observations to an absolute minimum* (and using a red light to preserve your dark adaptation as you do). Some observers spend far more time writing about what they see at the eyepiece than they actually do viewing it!

Finding Them

Here we examine the crucial matter of locating the objects in the Herschel catalog. One way is the use of traditional, mechanical or digital setting circles on your telescope's mounting to "dial up" each object. There is also the much more modern Go-To or GPS technology, which uses a computerized mounting to essentially automatically find (and track) targets. The two leading telescope companies supplying this technology are Celestron and Meade. The databases in even their basic Go-To instruments contain thousands of objects and the number swells to more than 14,000 in their premium models. Among these are many Herschel objects – especially the 615 prominent ones I proposed as a goal for Herschel clubs (and which are listed in Appendix 3 of this book). While some of them can actually be pulled up on the telescope's keypad by entering their common name (the "Blinking Planetary" or "Jupiter's Ghost," for example), most require knowing the object's *NGC* number. In no case can the original Herschel designations be used to access targets, these being considered outmoded or obsolete by the various programmers of these systems. We can only wonder what Sir William would think of this assessment!

Even using these modern devices to "navigate" your telescope for you, some knowledge of at least where the brighter stars are located is needed if for no other reason than providing initial alignment for their computers. And here the author wishes to emphatically state that the real fun of leisurely stargazing (at least for many of us purists!) is traditional "star-hopping" to desired targets using a good finder and the surrounding star patterns to locate them. (And quite often, the sights encountered along the way are as fascinating as the objects being sought after. So even if the telescope you already own or are planning to buy offers automated finding, you are encouraged to still spend at least some time star-hopping your way around the majestic highways and byways of the night sky. You will be glad you did!) (Fig. 4.6).

This, of course, requires a star atlas. Such widely used ones as *Norton's Star Atlas*, *The Cambridge Star Atlas*, and the *Bright Star Atlas 2000.0* are fine for general navigation of the heavens, but successful star-hopping to the sky's fainter denizens requires a more detailed celestial roadmap – one showing stars to at least as faint as 8th-magnitude, or about the limit of most finders and binoculars. While there are several highly detailed, multi-volume atlases readily available today

Fig. 4.6. Some popular references used for finding deep-sky wonders, including the Herschel objects themselves: left, *Norton's Star Atlas*; middle, *Sky & Telescope's Pocket Sky* Atlas; right The *Cambridge Star Atlas*. Photo by Sharon Mullaney.

(such as *Uranometria 2000.0*, which plots over 280,000 stars to magnitude 9.75 and shows more than 30,000 nonstellar objects – including *all* of William and John Herschel's discoveries!), a more practical one for deep-sky observers is *Sky Atlas 2000.0* (Sky Publishing and Cambridge University Press, 1998). Its more than 81,000 stars of visual magnitude 8.5 and brighter (plus some 2,700 nonstellar wonders) provide an ideal number of "signposts" for effective star-hopping to targets employing typical finders without resulting in "overkill." All 615 Herschel objects listed in Appendix 3 are plotted on its maps, as well as many of those in Classes II and III. Unfortunately, none of them are designated by their Herschel classes and numbers but rather by their *NGC* numbers. The same is true for *Uranometrica 2000.0* and all other modern star atlases.

This sad state of affairs (at least for Herschel enthusiasts) brings us back to *Norton's Star Atlas*. There are actually several different "Nortons" – as this classic work is widely referred to by observers. First, there is the original *Norton's Star Atlas*, first published in 1910 by Arthur Norton himself. This ran through 17 editions up into the late 1980s, in which the handbook section was continually updated by Norton and others after his death, but the star maps themselves were left untouched. Then in 1989 came an 18th (followed by a 19th) edition known as *Norton's 2000.0* featuring totally redrawn star maps and an extensively revised handbook. Finally, 2004 saw another completely revised work with yet another set of new star maps, this 20th edition reverting back to the original title of *Norton's Star Atlas*.

The original *Norton's* (and all subsequent editions of it up through the 17th) used Herschel's designations in plotting deep-sky objects on the star maps in addition to those of Messier, and *NGC* ones at low declinations. But, regrettably, the later new editions dropped the Herschel classes and numbers in redrawing their star maps (as well as all of the original double star discoverers' designations for those objects!). Thus, observers desiring to see the Herschel objects plotted on a star atlas will find them only on the maps of those editions of *Norton's* from the 17th or earlier. (The total number shown is 317, which include many of his showpieces.)

The real problem here is *finding* any of those treasured earlier editions! And as an update here, *Sky & Telescope's* Roger Sinnott recently published his superbly detailed and executed *Pocket Sky Atlas* (Sky Publishing, 2006). This spiral-bound work *does* specifically plot the original Herschel Club-400 target list (and many other Herschel objects as well) but, unfortunately, it uses their *NGC* designations rather than Sir William's.

Personal Matters

There are a number of little-recognized factors that impact the overall success of an observing session at the telescope. One concern is dressing properly. This is of particular importance in the cold winter months of the year, when observers often experience sub-freezing temperatures at night. It is impossible to be effective at the eyepiece – or even to just enjoy the views – when you are numb and half-frozen to death! Proper protection of the head, hands and feet are especially critical during such times and several layers of clothing are recommended as opposed to one heavy one for thermal insulation of the body in general. (It is said that William Herschel rubbed raw onion on his face as protection from the cold. He often worked on nights so frigid that ink froze in its well *inside* the upstairs room of his house where his sister Caroline took his dictation through an open window, called to her from the telescope!) During the summer months, the opposite problem arises, as observers attempt to stay cool and dry. In addition to very short nights at this time of the year, there is the added annoyance of flying insects, together with optics-fogging humidity and dew.

Another concern is proper posture at the telescope. It has been repeatedly shown that the eye sees more in a comfortably seated position than when standing, twisting, or bending at the eyepiece! If you must stand, be sure that the eyepiece/focuser is at a position where you do not have to turn and strain your neck, head, or back to look into it. This is especially important in using large reflectors, which often require a ladder just to reach the eyepiece! And while not as critical, the same goes for positioning finder scopes where they can be used without undue contortions.

Proper rest and diet also play a role in experiencing a pleasurable observing session. Attempting to stargaze when you are physically and/or mentally exhausted is guaranteed to leave you not only frustrated – but also looking for a buyer for your prized telescope! Even a brief "cat nap" before going out to observe after a hectic day is a big help here. Heavy dinners can leave you feeling sluggish and unable to function alertly at the telescope. It is much better to eat after stargazing – especially so, since most observers find themselves famished then (particularly on cold nights!). Various liquid refreshments such as tea, coffee, and hot chocolate can provide a needed energy boost (and warmth when desired). And while alcoholic drinks like wine do dilate the pupils, technically letting in more light, they also adversely affect the chemistry of the eye. This reduces its ability to see fine detail on the Moon and planets or resolve close double stars, and especially its sensitivity in viewing "faint fuzzies" like dim nebulae and galaxies!

Finally, there is the very important matter of *preparation*. And here, it is not just being aware of what objects are visible on a given night at a particular time of year, or which of them can be seen from your site and with your instrument, or deciding on those you plan to observe this time out. It goes beyond this to understanding something about the physical nature of the wonders you are looking at – be it a star cluster or nebula or remote galaxy – and its place in the grand cosmic scheme of things. In other words, as stargazers we must "see" with our minds as well as our sight. To better appreciate the importance of preparation before going to the telescope, the author would like to share with you the following lines from Charles Edward Barns' long out-of-print classic *1001 Celestial Wonders*:

> Let me learn all that is known of them,
> Love them for the joy of loving,
> For, as a traveler in far countries
> Brings back only what he takes,
> So shall the scope of my foreknowledge
> Measure the depth of their profit and charm to me.

Part II

Exploring The Herschel Showpieces

Chapter 5

Showpieces of Class I

Bright Nebulae

Listed below in alphabetical order by constellation are 57 of the most interesting objects in Herschel's Class I. Following the Herschel designation itself is the corresponding *NGC* number in parentheses, its Right Ascension and Declination (for Epoch 2000.0), the object's actual type (which may differ from the Class Herschel assigned it to), its visual magnitude, angular size in minutes (′) or seconds (″) of arc, and Messier or Caldwell number plus popular name if any. Next is a translation of Sir William's shorthand description (in italicized quotes) taken from the *NGC* itself, followed by comments from the author. These include directions for finding each object by sweeping for it, just as Herschel himself originally did.

Aries

H I-112 (NGC 772): 01 59 + 19 01, galaxy, 10.3, 7′ × 4′. *"Bright, considerably large, round, gradually brighter in the middle, resolvable (mottled, not resolved)."* Positioned only about 1.5° E of the beautiful double star γ Arietis, this spiral would seem to be a snap to locate. But in 3- to 6-inch telescopes, the author has always found this to be a difficult object to see. Its elongated asymmetric shape and brighter center can be glimpsed in 10-inch and larger apertures. Note that Herschel described it as resolvable in his large reflectors, as he did for many other galaxies. While he could not possibly have seen individual stars on the verge of resolution in these remote objects, he may have detected clumps of stars and bright nebulosity within them.

Bootes

H I-34 (NGC 5248): 13 38 + 08 53, galaxy, 10.3, 6′ × 5′ = Caldwell 45. *"Bright, large, extended, 150 degrees, pretty suddenly brighter in the middle to a resolvable (mottled, not resolved) nucleus."* Here is a fairly big oval-shaped spiral galaxy with bright nucleus showing hints of its two large, graceful spiral arms in 12- to 14-inch apertures. Situated all by itself near the SW corner of Bootes at the Virgo border, it can be readily picked up by careful sweeping.

Cancer

H I-2 (NGC 2775): 09 10 + 07 02, galaxy, 10.3, 4′, = Caldwell 48. "*Considerably bright, considerably large, round, very gradually then very suddenly much brighter in the middle, resolvable (mottled, not resolved).*" This spiral galaxy appears as a perfectly round glow with a noticeably brighter center. Located just over the border of Cancer from Hydra, NE of the Sea Serpent's "head," its isolation from any other bright galaxies in the area makes it easy to sweep up and identify.

Canes Venatici

H I-195 (NGC 4111): 12 07 + 43 04, galaxy, 10.8, 5′ × 1′. "*Very bright, pretty small, much extended 151 degrees.*" Sitting right on the border of Canes Venatici and Ursa Major, this elliptical galaxy is sometimes found listed under the latter constellation. Noticeably egg-shaped in outline, some observers report seeing a star-like nucleus here. Interestingly, Herschel did not mention it. The dim 11th-magnitude elliptical galaxy H IV-54 (NGC 4143) lies less than 1°S, within the same wide eyepiece field of view.

H I-213 (NGC 4449): 12 28 + 44 06, galaxy, 9.4, 5′ × 4′, = Caldwell 21. "*Very bright, considerably large, much extended, double or bifurcated, well resolved, clearly consisting of stars, a star of 9th magnitude following 5′.*" Here is a nice, bright irregular galaxy having a strange rectangular or box-shape to it as seen in larger apertures. Herschel implied this odd shape in his "bifurcated" remark. Even more interesting is the fact that he felt it was clearly resolved into stars – something possible for a nearby irregular system like this in really large apertures. You will find it midway between the previous galaxy (H I-195) and the variable star Y Canum Venaticorum, better known as La Superba from its lovely ruddy-orange hue.

H I-198 (NGC 4490): 12 31 + 41 38, galaxy, 9.8, 6′ × 3′, Cocoon Galaxy. "*Very bright, very large, much extended 130 degrees, resolvable (mottled, not resolved), south following of two.*" This spiral appears as a large, uniformly bright cosmic egg in a 6-inch glass. But how the name "Cocoon" applies to it is somewhat of a puzzle to the author. Take a look at it for yourself and see if you can tell why? The other object Herschel referred to is his H I-197 (NGC 4485), an 11th-magnitude elliptical galaxy lying just 3′ N within the same eyepiece field. Both objects are located just 1°NW of β Canum Venaticorum, making them easy to find (Fig. 5.1).

H I-176/177 (NGC 4656/7): 12 44 + 32 10, galaxy, 10.4, 14′ × 3′, Hockey Stick Galaxy. "*Remarkable, pretty bright, large, very much extended 34 degrees, south preceding of two.*" / "*Remarkable, pretty faint, large, extended 90 degrees +/−, north following of two.*" The primary object here is an irregular system appearing as a very long and narrow ray, from one end of which extends a fainter, tiny companion 1′ in size curving out at nearly a right angle. The overall appearance is indeed that of a hockey stick – especially as seen in 8-inch and larger apertures. Many references, including *Sky Catalogue 2000.0*, combine this pair into a single object, despite its having two Herschel/*NGC* numbers, so close together are they. Located just above the Coma Berenices border, about 30′ SE of the striking Humpback Whale Galaxy (placed in Class V by Herschel – see Chapter 7), this unequal duo lies in the same low-power eyepiece field with it. The distance of the Hockey Stick is 25,000,000 light-years (Fig. 5.2).

Fig. 5.1. H I-198 (NGC 4490), better-known as the Cocoon Galaxy, is a big graceful spiral easily seen in small apertures. The tiny elliptical galaxy H I-197 (NGC 4485) lies in the same eyepiece field, but at 11th-magnitude it is not an easy catch. Courtesy of Mike Inglis.

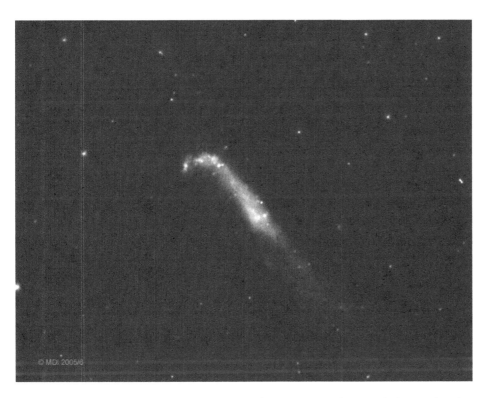

Fig. 5.2. H I-176/177 (NGC 4656/7) is a pair of galaxies – one large and obvious, the other small and elusive – apparently attached to each other in such a way that observers have likened their unusual shape to that of a hockey stick as seen in larger apertures. Courtesy of Mike Inglis.

Fig. 5.3. H I-186 (NGC 5195) is the companion to the Whirlpool Galaxy (M51). Although Messier no doubt saw both objects, his number 51 refers to the large spiral itself – allowing William Herschel to add the small galaxy to his catalog. Courtesy of Mike Inglis.

H I-96 (NGC 5005): 13 11 + 37 03, galaxy, 9.8, 5′ × 3′, = Caldwell 29. "*Very bright, very large, very much extended 66 degrees, very suddenly brighter in the middle to a nucleus.*" This bright spiral is located SE of the beautiful double star Cor Caroli (α Canum Venaticorum), just midway between the stars 14 and 15/17 Canum Venaticorum. Its elliptical shape and bright core are obvious even in a small glass, while hints of its tightly wound spiral arms can be seen in large aperture instruments.

H I-186 (NGC 5195): 13 30 + 47 16, galaxy, 9.6, 5′ × 4′, Companion to Messier 51. "*Bright, pretty small, little extended, very gradually brighter in the middle, involved in M51.*" Seemingly attached to the end of one of the spiral arms of the famed Whirlpool Galaxy (giving it the aspect of a double "nebula"), this peculiar irregular system is one virtually every deep-sky observer has seen but few realized it is a Herschel object (and one of the easiest to find in the sky). The big Whirlpool and its companion form an equilateral triangle with η Ursae Majoris (the star at the end of the Big Dipper's handle) and 24 Canum Venaticorum, and are easily spotted even in a 3- or 4-inch glass. H I-186 itself looks nearly round with a brighter center. While it took Lord Rosse's mammoth 72-inch metal-mirrored reflector to initially see them, the spiral arms of M51 can actually be glimpsed in an 8-inch on a dark night. This provides a dramatic illustration of Herschel's well-known dictum that "When an object is once discovered by a superior power, an inferior one will suffice to see it afterwards"! The graceful Whirlpool lies some 31,000,000 light-years from us (Fig. 5.3).

Coma Berenices

H I-19 (NGC 4147): 12 10 + 18 33, globular cluster, 10.3, 4′. "*Globular cluster of stars, very bright, pretty large, round, gradually brighter in the middle, well resolved, clearly consisting of stars.*" Located midway between the stars 3 and 5 Comae

Berenices, near the W edge of this galaxy-strewn constellation, we find this tiny and highly compact starball. In small scopes it looks like a little round nebula with a brighter core, and it really takes a good 12- to 14-inch on a steady night to make this object look like a globular cluster. The puzzle here is why Herschel assigned it to his Class I (bright nebulae) since he stated that it was well resolved and clearly consisted of stars! It would seem to belong in his Class VI (very compressed and rich clusters of stars), which contains many other globulars.

H I-75 (NGC 4274): 12 20 + 29 37, galaxy, 10.4, 7′ × 3′. *"Very bright, very large, extended 90 degrees, much brighter in the middle to a nucleus."* This big elliptical-shaped object has been reported to have a Saturn-like inner ring from its appearance on short-exposure images. Visually, it shows a bright central region surrounded by a faint outer halo. And perhaps the ring structure can be seen in large amateur scopes – but it cannot be very obvious, for Herschel himself made no mention of it. (But again, now knowing that it is there may make a difference!) This spiral is located in a galaxy-rich area just above the big naked-eye Coma Star Cluster, NW of the star γ Comae Berencies at the cluster's N edge. Lying as it does nearly on the +30° Declination line, careful sweeping along it will bring it into view.

H I-92 (NGC 4559): 12 36 + 27 58, galaxy, 10.0, 11′ × 5′, = Caldwell 36. *"Very bright, very large, much extended 150 degrees, gradually brighter in the middle, three stars following."* This big multiple-arm spiral can be found by sweeping about 2°E of γ Comae Berenices, the N-most and brightest star in the Coma Star Cluster. The arms themselves may not be directly visible even in large apertures but a kind of "lumpiness" may be evident indicating the presence of unresolved arms and the dust lanes between them. In telescopes 6 inches or less in size, this object looks neatly elongated and uniform in brightness across the image with no obvious nucleus.

H I-84 (NGC 4725): 12 50 + 25 30, galaxy, 9.2, 11′ × 8′. *"Very bright, very large, extended, very gradually,[then] very suddenly very much brighter in the middle to an extremely bright nucleus."* This spiral's arms are so long and completely wrapped around its central hub that it has a ring-like appearance on photographs. Aside from its obvious elliptical shape, visually it shows a bright star-like nucleus. It is located about 2°S and slightly W of the star 31 Comae Berenices (itself situated right at the North Galactic Pole).

Corvus

H I-65 (NGC 4361): 12 24 − 18 48, planetary nebula, 10.3, 80",= Caldwell 32. *"Very bright, large, round, very suddenly much brighter in the middle to a nucleus, resolvable (mottled, not resolved)."* Located within the "sail" of Corvus – near the intersection of the diagonals drawn from its four corner stars – this big round planetary would seem to be an easy catch. But the author has always found it elusive in small apertures unless skies are really dark. Part of the reason is its large apparent size, resulting in a low surface brightness (which is only about a quarter that of the famed Ring Nebula, M57, in Lyra). Under good conditions, it can be seen in a 4-inch glass and there are even reports of it having been spotted in large binoculars. Its very faint 13th-magnitude central star has been glimpsed in telescopes as small as an 8-inch and some observers have reported that its disk is filled with a

Fig. 5.4. The big and somewhat elusive planetary nebula, H I-65 (NGC 4631). It looks more like a dim galaxy seen face-on in the eyepiece than a typical planetary, likely accounting for why Sir William did not assign it to the latter class. Courtesy of Mike Inglis.

mottled light – both features mentioned in Sir William's description. This unusual planetary is one of the better-known examples of a Herschel object being assigned to the wrong class. But in all fairness to Sir William, it could easily pass for a small galaxy (nebula). This dim sphere is located 2,600 light-years from us (Fig. 5.4).

Delphinus

H I-103 (NGC 6934): 20 34 + 07 24, globular cluster, 8.8, 7′, = Caldwell 47. *"Globular cluster of stars, bright, large, round, well resolved, clearly consisting of stars, stars from the 16th magnitude downwards, star of 9th magnitude preceding."* This compact starball is the brighter of the two small globulars located within Delphinus. Sweep for it about 4° due S of ε Delphini. Although Herschel called it bright, large and well-resolved, you will need at least a 12-inch scope to see it that way. In small instruments, it appears as a fuzzy ball with a brighter center. There is also a fairly prominent field star near it which will be found very useful in achieving a sharp focus for those attempting to resolve the cluster itself. Here is another case where you would think Herschel should have placed this object in his Class VI since he clearly resolved it into a cluster of stars.

H I-52 (NGC 7006): 21 02 + 16 11, globular cluster, 10.6, 4′, = Caldwell 42. *"Bright, pretty large, round, gradually brighter in the middle."* The second globular here lies about 4° due E of γ Delphini – the star marking the tip of the Dolphin's nose (and itself a truly magnificent yellow and green double star which should not be missed). This object looks like a fainter, smaller version of H I-103, but in addition it has a definite "remote" look to it – which is interesting because it is, in fact, one of the most distant objects of its type. While it can be seen in a 6-inch, like its neighbor it needs an aperture at least twice that size to see it as Herschel did. And since he apparently did not resolve it, in this case we can understand why it was assigned to Class I as a "bright nebula" even though it is actually a cluster. The only real question here (at least in the author's mind) is how this object ever merited a place in the Caldwell listing?

Draco

H I-215 (NGC 5866): 15 07 + 55 46, galaxy, 10.0, 5′ × 2′, Formerly Messier 102. "*Very bright, considerably large, pretty much extended 146 degrees, gradually brighter in the middle.*" Here is one famous Herschel object that has sadly lost its claim to fame! Long believed to be the 102nd entry in Messier's famous catalog, more recent historical study has shown this to have been a duplication of M101 in neighboring Ursa Major, resulting in its removal from his list. This big elliptical system is located about 4°SW of ι Draconis and is part of a trio of Herschel galaxies situated at the Draco-Bootes border. (The second of these is the Splinter Galaxy, H II-759/NGC 5907, described in Chapter 11 and the third is H II-757/NGC 5879, an 11.5-magnitude spiral.) This cosmic egg is easy to sweep up even in a 4-inch glass, but some care is needed to make sure it is the right galaxy! All three objects can be seen within the same low-power eyepiece field in a 6-inch rich-field telescope (RFT) on dark, transparent nights (Fig. 5.5).

Fig. 5.5. Traditionally, H I-215 (NGC 5866) had been included in the Messier list as M102. But it has since been found that this was simply a duplication of M101. What a pity, for this would surely have been a very worthy entry in that famous roster. Courtesy of Mike Inglis.

Eridanus

H I-64 (NGC 1084): 02 46 – 07 35, galaxy, 10.6, 3′ × 2′. *"Very bright, pretty large, extended, gradually pretty much brighter in the middle."* This little spiral sits on the border of Eridanus and Cetus, about 4°NW of the star η Eridani. Appearing slightly egg-shaped with a brighter center, it is typical of the many members of its class scattered throughout these adjoining constellations. In fact, it lies just 1.5°NE of the galaxy pair NGC 1042 and NGC 1052 in Cetus. (The latter is H I-63 but the former was apparently not seen by Herschel, even though the two objects lie close together in the same eyepiece field.)

H I-107 (NGC 1407): 03 40 – 18 35, galaxy, 9.8, 2′. *"Very bright, large, round, suddenly very much brighter in the middle to a nucleus."* You will find this bright but tiny elliptical by sweeping about 1.5°SE of 20 Eridani. At low magnification in a 4- to 6-inch glass it looks like a slightly enlarged star, while higher powers show a fuzzy edge to it. Just 12′ SW lies the dim spiral NGC 1400, which is H II-593. With a magnitude variously listed as 11th to fainter than 12th, it requires a dark night and the use of averted vision to glimpse in 8- to 10-inch apertures.

Leo

H I-56/57 (NGC 2903/5): 09 32 + 21 30, galaxy, 8.9, 13′ × 7′. *"Considerably bright, very large, extended, gradually much brighter in the middle, resolvable (mottled, not resolved), south preceding of two."/"Very faint, considerably large, round, pretty suddenly brighter in the middle, resolvable (mottled, not resolved), north following of two."* Here is a fascinating spiral that was cataloged as two separate objects by Sir William but which in reality is only one! It is also one of the brightest galaxies missed by Messier and one of the finest in the entire NGC. Not only is it big and luminous, but it is easily found floating all alone about 1.5° due S of λ Leonis – itself just W of the top of the Sickle asterism marking the Lion's head. And part of this object's appeal is its isolation from the confusing swarms of galaxies both in Leo itself and in the vast Coma-Virgo Cluster to the E, making it easy to locate and identify. H I-56 is the actual galaxy itself while H I-57 is a dim starcloud lying 12′ N of the nucleus, giving this object the appearance of a double "nebula" (or having a double nucleus) in very large apertures. But the latter is a difficult catch in amateur-class instruments, buried as it is in the glow of its parent galaxy. This misty jewel lies some 30,000,000 light-years from us, yet it is bright enough to be seen even in a 3-inch glass on a dark night (Fig. 5.6).

H I-17 (NGC 3379): 10 48 + 12 35, galaxy, 9.3, 4′, = Messier 105 *"Very bright, considerably large, round, pretty suddenly brighter in the middle, resolvable (mottled, not resolved)."* This bright spiral lies in the same low-power field with M95 and M96. And with the 10th magnitude 6′ × 3′ elliptical H I-18 (NGC 3384) nearly on top of H I-17 itself, this group offers a fascinating sight for galaxy lovers. (There is even the dim 12th-magnitude elliptical galaxy H II-41/NGC 3389 lurking in the field, forming a tight triangle with H I-17 and H I-18.) You can readily find this fascinating clan by sweeping midway between the stars 52 and 53 Leonis, which are themselves located E of Regulus (α Leonis). An 8-inch telescope equipped with a wide-angle eyepiece neatly shows this group as a small cluster of galaxies, shining

Fig. 5.6. This bright spiral has the double Herschel designation H I-56/57 (NGC 2903/5), but is actually just a single object with a nebulous patch seen near its nucleus in large telescopes. Messier somehow missed this galaxy, leaving it to Sir William to discover this showpiece. Courtesy of Mike Inglis.

from across some 30,000,000 light-years of intergalactic space. Circular H I-17 itself reminds some observers of an unresolved globular with bright core. And just for the record, Sir William's description of H I-18 reads: "*Very bright, large, round, pretty suddenly much brighter in the middle, 2nd of 3.*"(Fig. 5.7).

H I-13 (NGC 3521): 11 06 – 02, galaxy, 8.9, 10′ × 5′. "*Considerably bright, considerably large, much extended 140 degrees +/–, very suddenly much brighter in the middle to a nucleus.*" This big, bright, multiple-arm spiral floats in isolated splendor just 30′ E of 62 Leonis. Its elongated shape and luminous core are obvious in scopes as small as a 4-inch. And yet this neglected object seems little known to deep-sky observers, apparently being overshadowed by Leo's Messier galaxies.

Lynx

HI-218 (NGC 2419): 07 38 + 38 53, globular cluster, 10.4, 4′, = Caldwell 25/ Intergalactic Wanderer. "*Pretty bright, pretty large, little extended 90 degrees, very gradually brighter in the middle, a star of 7.8 magnitude at 267 degrees, 4′ distant.*" This amazing object is famed as being the most remote globular cluster known (other than those in other galaxies), lying 300,000 light-years from us far beyond the edge of our Milky Way. Indeed, it is speculated that this tiny starball may actually be an intergalactic object – wandering between the galaxies of our Local Group. You will find it 7°N of the bright binary star Castor (α Geminorum) in

Fig. 5.7. H I-17 (M105) is the smaller of the three spiral galaxies shown here, the other two being M 95 and M96. This striking trio is an easy catch even in small glasses. The fainter elliptical HI-18 (NGC 3384) lies nearly on top of H I-17 itself. Courtesy of Mike Inglis.

neighboring Gemini. Two fairly prominent stars (one of which is mentioned in Herschel's description) lie in the same eyepiece field pointing right at H I-218, the space between them and the cluster being roughly equal. A 5-inch scope shows it as a fuzzy ball with a brighter center. On nights of good seeing (image steadiness), hints of resolution (or "graininess") begin to appear in apertures twice that size and in a 14-inch it actually sparkles! However, Herschel said nothing at all about being able to resolve it in his large 20-foot (18.7-inch) reflector and considered this to be a nebula. Poor seeing and/or the cooling and changing figure of his speculum-metal mirror may have been at play here (Fig. 5.8).

H I-200 (NGC 2683): 08 53 + 33 25, galaxy, 9.7, 9′ × 2′. "*Very bright, very large, very much extended 39 degrees, gradually much brighter in the middle.*" This lovely edge-on spiral is located about 1°NW of neighboring σ-2 Cancri (one of a little group of four naked-eye stars carrying the designation σ). This is among the brighter springtime galaxies, yet it is largely unknown to most deep-sky observers. What a pity! A 4-inch glass shows it nicely but an 8-inch is needed to fully appreciate the view, in which its distinct spindle or cigar-shape and bright center stand out prominently. We can only wonder what Sir William's reaction must have been upon coming across objects like this after encountering hosts of fainter denizens of deep space in his surveys of the sky. It surely must have spurred him on, thinking about what other unsuspected wonders might lay ahead in the very next sweep!

Ophiuchus

H I-48 (NGC 6356): 17 24 – 17 49, globular cluster, 8.4, 7′. "*Globular cluster of stars, very bright, considerably large, very gradually very much brighter in the middle, well resolved, clearly consisting of stars, stars 20th magnitude.*" This little starball

Fig. 5.8. HI-218 (NGC 2419) is actually a very remote globular cluster rather than a nebula as Herschel cataloged it. It has been dubbed the Intergalactic Wanderer due to its vast distance (for a globular), actually lying outside the confines of the Milky Way itself in intergalactic space! Courtesy of Mike Inglis.

was chosen as being typical of the many non-Messier globulars Herschel discovered in this expansive constellation (it actually being one of the brightest of those he found). It lies 4°SE of η Ophiuchi and just over 1°NE of the well-known globular M9 (itself the smallest of the Messier globulars here with an apparent diameter of just 9′). It requires a good 8- or 10-inch and fairly high magnification on a steady night to resolve this tight stellar beehive. As we have asked in similar instances, why did Herschel place this object in his Class I when he was able to clearly resolve it into a cluster of stars? (But here as elsewhere, see the important note at the end of Chapter 3.)

Pegasus

H I-53 (NGC 7331): 22 37 + 34 25, **galaxy, 9.5, 11′ × 4′, = Caldwell 30.** *"Bright, pretty large, pretty much extended 163 degrees, suddenly much brighter in the middle."* Here is yet another of the best galaxies overlooked by Messier – this one a big, nearly edge-on spiral that reminds some observers of a smaller version of the Andromeda Galaxy (M31). It appears as an elongated smudge in 3- to 6-inch apertures while 10- to 12-inch scopes reveal traces of the nuclear bulge and spiral arms (actually the dust lanes between them). Sweep for it about 4°N and slightly W of η Pegasi. We view this graceful pinwheel from across a distance of some 50,000,000 light-years. Lying in the same eyepiece field – just 30′ to the SSW – is the famed and often-pictured cluster of galaxies known as Stephan's Quintet. Visually a difficult catch, its members range in brightness (actually dimness!) from 13th- to 14th-magnitude (some sources listing them as a full magnitude fainter).

Fig. 5.9. H I-53 (NGC 7331) is another fine galaxy missed by Messier. This spiral is oriented N-S and tilted some 15° to our line of sight. While there are other faint galaxies nearby, it stands out all along as seen in small instruments. Courtesy of Mike Inglis.

Interestingly, Herschel apparently saw none of its members, attesting to the difficulty of observing this cluster visually. But here again, knowing of its existence may make a difference. So readers with large telescopes looking for a challenge should give it a try (Fig. 5.9).

H I-55 (NGC 7479): 23 05 + 12 19, galaxy, 11.0, 4′ × 3′, = Caldwell 44. *"Pretty bright, considerable large, much extended 12 degrees, between two stars."* This rather dim S-shaped barred spiral is located about 3°S of α Pagasi. At least an 8-inch aperture will be needed to clearly see its nucleus and faint outer regions. Those blessed with telescopes in the 12- to 14-inch range will also be able to make out its two big spiral arms protruding from the central bar. There are a couple of faintish stars surrounding this object, two of which are mentioned in Herschel's description. This gossamer pinwheel definitely seems more like it belongs in Class II rather than in Class I to the author. What is your impression?

Perseus

H I-193 (NGC 650/1): 01 42 + 51 34, planetary nebula, 11.5, 140″ × 70″, = Messier 76/Little Dumbbell/Barbell/Cork/Butterfly Nebula. *"Very bright, following of a double nebula."* This multiple-named wonder is widely considered to be the faintest of the Messier objects and one of the most difficult to see. And yet it is visible in a 3-inch glass on a dark night and is a fascinating sight in 6-inch and larger scopes. Sweep for it just 1°NW of φ Persei, which sits right on the border with neighboring Andromeda. This curious object has been variously described by observers as having a rectangular, cork or peanut shape, as a double-lobed nebula, and appearing like a miniature of the famed Dumbbell Nebula (M27) in Vulpecula. And, indeed, it is all of these and more – depending on the aperture used. Note in particular Herschel's comment about a "double nebula." NGC 650 is M76 itself, which he obviously recognized as such and thus it carries no H-designation. H I-193 (NGC 651) refers to the fainter half of the nebula, which Sir William apparently considered to be a second independent object here and so included it in his catalog. This unusual planetary lies in a rich region of the Milky Way and seems to "float" against the starry background. And while some observers report that the nebula has a bluish-green hue to it (which is indeed very typical of the brighter planetaries), the author has never really seen any definite color here. Like most objects of its class, its distance is not precisely known – in this case values ranging all the way from 2,000 to 4,000 light-years are given by various sources (Fig. 5.10).

H I-156 (NGC 1023): 02 40 + 39 04, galaxy, 9.5, 9′ × 3′. *"Very bright, very large, very much extended, very very much brighter in the middle."* This peculiar elliptical system lies just over 1°S of the star 12 Persei, itself located W of the famed eclipsing binary Algol (β Persei). Here we find an obvious egg-shaped glow with a bright nucleus. There are few bright galaxies to be found in or near the Milky Way itself, but this one happens to lie right at its outer fringe and as a result appears suspended amidst a fairly rich strata of foreground stars.

Fig. 5.10. H I-193 (M76) is the well-known Little Dumbbell Nebula (among other names). It carries a dual NGC designation (NGC 650/1) due to its bipolar aspect. And while Herschel did see it as a double nebula, he assigned only a single number to it. This refers to its fainter section, the brighter one being what Messier had already cataloged. Courtesy of Mike Inglis.

Sagittarius

H I-150 (NGC 6440): 17 49 – 20 22, globular cluster, 9.7, 5′. *"Pretty bright, pretty large, round, brighter in the middle."* Here is a little globular that is quite typical of the many non-Messier ones populating this cluster-rich constellation. Sweep for it 2°SW of the big, beautiful open cluster M23. It lies about midway along a line connecting that glittering jewel box and the star 58 Ophiuchi. While a 6-inch clearly shows it as a small fuzzball with brighter center, twice that aperture is needed to begin to resolve it since the individual stars are quite faint and well compacted. Note that Herschel did not mention it being resolved or even mottled. Just 30′ N lies the 12th-magnitude planetary nebula H II-586 (NGC 6445), 35″ × 30″ in apparent size. Visible in a 6-inch telescope within the same wide eyepiece field as the globular itself, it reportedly has a bluish-green hue. It also seems to be brighter than the magnitude quoted here.

Scutum

H I-47 (NGC 6712): 18 53 – 08 42, globular cluster, 8.2, 7′. *"Globular cluster of stars, pretty bright, very large, irregular, very gradually a little brighter in the middle, well resolved, clearly consisting of stars."* This misty-looking starball lies 2.5° nearly due S of Scutum's famed Wild Duck Cluster (M11). As such, it is greatly overshadowed by its spectacular neighbor and is seldom observed. Although obvious in a 4-inch glass, it takes at least an 8-inch at high magnification and a steady night to resolve its outer stars. Situated as it is in a very rich part of the Milky Way known as the Scutum Star Cloud, it is a nice sight even if your telescope cannot show its individual members. Note that although Herschel clearly resolved it, he still placed it in his Class I. Some sources give its apparent size as small as 3′ in diameter, but it definitely looks bigger than this in the eyepiece. We view this stellar beehive from across a distance of 25,000 light-years. Adding to the interest here (at least for some observers!) is the dim 14th-magnitude planetary nebula IC 1295. Situated just 20′ to the ESE of H I-47 itself, this 80″ × 60″ spheroid is extremely difficult to glimpse in small instruments – something which the fact that it carries no Herschel designation and was not included in the *NGC* certainly supports.

Sextans

H I-163 (NGC 3115): 10 05 – 07 43, galaxy, 9.2, 8′ × 3′, = Caldwell 53/Spindle Galaxy. *"Very bright, large, very much extended 46 degrees, very gradually then suddenly much brighter in the middle extended to a nucleus."* Sweeping 1.5°NW of the close star-pair 17 and 18 Sextantis, or about 3°E and slightly N of γ Sextantis, bring us to this prominent elliptical galaxy. Easy to see in a 3-inch glass, this object is well-named with its spindle-shaped, elongated outline and bright center. Large amateur scopes reveal its pointed ends and may even occasionally show its star-like nucleus flashing – the latter likely the result of atmospheric turbulence. Widely looked upon as the sole deep-sky attraction of Sextans, H I-163 lies more than

20,000,000 light-years from us. Several other fainter Herschel galaxies actually do reside in this small constellation, including the pair H I-3 and H I-4 described below.

H I-3 (NGC 3166): 10 14 + 03 26, galaxy, 10.6, 5′ × 3′/H I-4 (NGC 3169): 10 14 + 03 28, galaxy, 10.5, 5′ × 3′. *"Bright, pretty small, round, pretty suddenly much brighter in the middle, second of three."/"Bright, pretty large, very little extended, pretty gradually much brighter in the middle, star of 11th magnitude at 78 degrees, 80″, third of three."* The first of the three objects Herschel mentions is NGC 3165 – which strangely does not have an H-designation even though it is stated as existing by him! (The NGC description of this 14th-magnitude 2′ × 1′ galaxy reads: "Very faint, much extended 0 degrees, first of three" and its actual position is 10 14 + 03 23.) H I-3 and H I-4 themselves lie nearly on top of each other, just 9′ apart. You will find them by sweeping about 1.5°S of 19 Sextantis. Both objects are slightly elliptical in shape and have fairly bright centers, but the main attraction here is their proximity to each other. Seen in an 8-inch or larger scope at low to medium magnifications, they present a fascinating sight!

Ursa Major

H I-205 (NGC 2841): 09 22 + 50 58, galaxy, 9.3, 8′ × 4′. *"Very bright, large, very much extended 151 degrees, very suddenly much brighter in the middle, equals a star of 10th magnitude."* This big bright spiral is another of the best of those missed by Messier, having been left to Herschel to discover. Look for it 2°SW of θ Ursae Majoris, the galaxy itself forming a triangle with that star and 15 Ursae Majoris. It is easily spied in 3- and 4-inch glasses, and is an impressive sight in scopes twice that big. They show it as a bright oval with an obvious star-like core. Photographs reveal a system of beautifully symmetrical spiral arms encompassing the nuclear hub, hints of which can be glimpsed in large apertures on a good night.

H I-168 (NGC 3184): 10 18 + 41 25, galaxy, 9.8, 7′. *"Pretty bright, very large, round, very gradually brighter in the middle."* Here is a fine face-on spiral located in the same low power eyepiece field as μ Ursae Majoris, lying about 30′ due W of the star itself. Circular in shape and very uniform-looking, this object is easily located thanks to μ. But it appears dimmer than you might expect since its light (which some sources give as fainter than 10th-magnitude) is spread out over a fairly large angular extent. The view here really does not seem to change much as telescope size increases, except for an overall boost in image brightness.

H I-201 (NGC 3877): 11 46 + 47 30, galaxy, 10.9, 5′ × 1′. *"Bright, large, much extended 37 degrees."* This edge-on spiral is among the fainter-looking Class-I objects but still well worth viewing due to the presence of χ Ursae Majoris close by in the field, just 16′ to the S. Not only does the star provide a striking contrast with the galaxy itself, but it makes this one of the easiest of all Herschel objects to find – simply center the star in the eyepiece and you are on the galaxy! It shows as a cigar-shaped streak in all apertures. (No mention of the star itself appears in Herschel's description, apparently due to the small fields of view of his telescopes.)

H I-203 (NGC 3938): 11 53 + 44 07, galaxy, 10.4, 5′. *"Bright, very large, round, brighter in the middle to a pretty bright nucleus, easily resolvable."* This face-on spiral is located about 1.5°NW of the star 67 Ursae Majoris – itself 9° due S of γ

Ursae Majoris in the bowl of the Big Dipper. An uniformly illuminated pale disk of light is what is typically seen in 4- to 8-inch telescopes. But note Herschel's comment that this object's nucleus was "easily resolvable" – perhaps he was seeing structure in its core (but certainly not stars).

H I-253 (NGC 4036): 12 01 + 61 54, galaxy, 10.6, 4′ × 2′/H I-252 (NGC 4041): galaxy, 11.1, 3′. "*Very bright, very large, extended.*"/"*Bright, considerably large, round, gradually, then pretty suddenly very much brighter in the middle to a resolvable (mottled, not resolved) nucleus.*" H I-253 is an egg-shaped elliptical with a bright nucleus, while H I-252 is a tiny round spiral also with a bright nucleus but noticeably fainter than its neighbor. This galaxy pair lies in the same eyepiece field just 16' apart and is a neat sight in 8-inch and larger scopes. Find them by sweeping at the apex of a triangle formed by α and δ Ursae Majoris, two of the Big Dipper's bowl stars. Interestingly, Herschel described 3′ H I-252 as "considerably large"??

H I-206 (NGC 4088): 12 06 + 50 33, galaxy, 10.5, 6′ × 2′/H I-224 (NGC 4085): galaxy, 12.3, 2′ × 1′. "*Bright, considerably large, extended 55 degrees, little brighter in the middle.*"/"*Bright, pretty large, pretty much extended 78 degrees, very suddenly brighter in the middle.*" Here is another unequal galaxy pair lying within the same eyepiece field. Both spirals, H I-206 is egg-shaped while just 11′ to the S is much smaller and elusive H I-224. Look for them in a field of galaxies 3°SE of γ Ursae Majoris in the Big Dipper's bowl. You probably will not notice H I-224 initially, for it takes attentive gazing to glimpse it. Apertures of 10-inch and larger are definitely needed to appreciate this contrasting duo. Although Herschel called it "bright," H I-224 is one of those objects that seems to belong more in Class II than Class I.

H I-231 (NGC 5473): 14 05 + 54 54, galaxy, 11.4, 3′ × 2′. "*Pretty bright, small, round, gradually brighter in the middle.*" Now here is a striking contrast in galaxies if ever there was one! Lying in the same wide eyepiece field on the outskirts of the big spiral M101 (the Pinwheel Galaxy) is this dim, miniscule elliptical just 30′ to its N. (M101 itself is located at the apex of an equilateral triangle formed by the naked-eye pair Mizar/Alcor at the bend of the Big Dipper's handle and η Ursae Majoris at the end of the handle.) Not only is H I-231 dim but it is also tiny, appearing both fainter and smaller than quoted here. (Indeed, some sources make it a full magnitude dimmer and less than 1′ in overall size.) In any case, it is definitely an elusive object in anything less than an 8-inch glass. It is actually just one of a field of faint galaxies that surrounds M101 itself. The view here on a dark, transparent night through a big, short-focus Dobsonian reflector equipped with a super-wide-angle eyepiece is one never to be forgotten.

Virgo

H I-9 (NGC 4179): 12 13 + 01 18, galaxy, 10.9, 4′ × 1′. "*Pretty bright, pretty small, pretty much extended 135 degrees +/–, brighter in the middle to a nucleus.*" About 1°SE of the star 10 Virginis we find an object which is variously classified as either an elliptical system or an edge-on spiral. In either case, it looks oval or spindle-shaped and displays a nucleus as seen in mid-sized instruments. It is quite typical of the hoards of galaxies scattered throughout this extensive constellation.

H I-35 (NGC 4216): 12 16 + 13 09, galaxy, 10.0, 8′ × 2′. "*Very bright, very large, very much extended 17 degrees, suddenly brighter in the middle to a nucleus.*" This lovely edge-on spiral sits right on the Virgo-Coma Berenices border, 2° due S of the star 6 in the latter constellation. Identification can be a real problem in this galaxy-rich area but this object's distinctive needle- or cigar-shape will aid in pin-pointing it. It is a nice sight in a 6-inch scope and a definite showpiece in 12-inch and larger apertures, which may also show the presence of two other extremely faint galaxies in the same eyepiece field.

H I-139 (NGC 4303): 12 22 + 04 28, galaxy, 9.7, 6′, = Messier 61. "*Very bright, very large, very suddenly brighter in the middle [like?] a star, binuclear.*" This is another of the very few Messier objects which Sir William included in his catalog even though he did not originally discover it. It is easily picked up by sweeping midway between the stars 16 and 17 Virginis (the latter being an unequal telescopic double). This big round spiral is a nice sight in medium-aperture scopes, while larger instruments show its bright star-like nucleus and some hint of its three prominent spiral arms. They may also reveal the bi-nuclear core seen by Herschel.

H I-28.1 (NGC 4435): 12 28 + 13 05, galaxy, 10.9, 3′ × 2′./H I-28.2 (NGC 4438): 12 28 + 13 01, galaxy, 10.1, 9′ × 4′, The "Eyes". "*Very bright, considerably large, round, north preceding of two.*"/"*Bright, considerably large, very little extended, resolvable (mottled, not resolved), south following of two.*" If you center your telescope on the two big, bright galaxies M84 and M86 at the core of the Virgo Galaxy Cluster and then sweep less than 30′ due E, you will come upon a pair of spirals that are so close to each other (just 4′ apart) that they remind many observers of a pair of eyes staring at them from across the depths of space!

Even the unusual Herschel designation suggests that they appear as two parts of the same object. Individually, H I-28.1 looks round and of uniform brightness while H I-28.2 is larger and obviously elongated with perhaps a trace of a nucleus. Some sources give their angular sizes as half those quoted here, better matching how they appear in the eyepiece. The Eyes are perhaps at their best in 6- to 8-inch telescopes, larger apertures tending to spread them too far apart and ruining the illusion (Fig. 5.11).

Fig. 5.11. H I-28.1/28.2 (NGC 4435/4438) is a close pair of spiral galaxies that look to some observers like eyes staring at them from across intergalactic space, leading to its popular name as The Eyes. They are rather faint as seen in small glasses, while really big scopes separate them unduly, ruining the effect. Medium apertures show the effect best. Courtesy of Mike Inglis.

H I-31 = H I-38 (NGC 4526): 12 34 + 07 42, galaxy, 9.6, 7′. "*Very bright, very large, much extended 120 degrees +/–, pretty suddenly much brighter in the middle, between two stars of 7th magnitude.*" This elliptical was apparently cataloged twice by Herschel, resulting in it having a dual designation. It is located about 1°SE of another big bright member of the Virgo Galaxy Cluster, M49 – itself lying S of the Cluster's core. Care is needed in pinpointing it, for there are other galaxies surrounding it. Fortunately, as indicated in Herschel's description, it happens to lie between two 7th-magnitude foreground stars, helping to distinguish it from its neighbors. It is noticeably egg-shaped with a fairly bright nucleus as seen in an 8-inch telescope.

H I-24 (NGC 4596): 12 40 + 10 11, galaxy, 10.5, 4′ × 3′. "*Bright, pretty small, round, gradually much brighter in the middle, resolvable (mottled, not resolved), three stars following.*" Positioned 30′ due W of the star ρ Virginis lies this barred spiral, appearing round with a very obvious sharp nucleus as seen in small scopes. Instruments in the 12- to 14-inch aperture range may allow glimpses of a Saturn-like disk with projecting ansae that appear on photographs, although Herschel himself made no mention of these features. In any case, look for the dim 11th-magnitude spiral galaxy H II-69 (NGC 4608) lying in the same wide eyepiece field, just 11′ SW of ρ.

H I-43 (NGC 4594): 12 40 – 11 37, galaxy, 8.3, 9′ × 4′, = Messier 104/Sombrero Galaxy. "*Remarkable, very bright, very large, extremely extended 92 degrees, very suddenly much brighter in the middle to a nucleus.*" Why Messier never added this marvelous object to his original catalog is a mystery, for it is one of the brightest and finest edge-on spiral galaxies in the entire sky! It shows up in even a 2-inch glass, and can actually be glimpsed in binoculars and finders on a dark night. In fact, it is so bright that it can be seen in full moonlight or through moderate light pollution in 3- and 4-inch apertures. And this, despite its distance of 28,000,000 light-years! Sweep for it at the base of a right triangle formed by Spica (α Virginis) and γ Virginis (better known as the marvelous but currently snug binary Porrima), sitting directly on the Virgo–Corvus border. Finding it is aided by two amazing asterisms that guide the observer right to it. Just 20' WNW from the Sombrero, within the same eyepiece field, is an arrow-shaped group of stars (the multiple system Struve 1664) that actually points right at the galaxy itself. This asterism has been dubbed "Little Sagitta" from its resemblance to that constellation, as well as the "Arrow." A beacon to the arrow lies to its SW, just over the border in Corvus. This is a striking and unique triangle of stars within a starry triangle that has been dubbed the "Stargate" and also the "Wedge." Both asterisms are clearly visible in a finder on all but the worst of nights, making it a cinch to find this stunning galaxy even for novice observers. A 6-inch telescope shows a huge bulbous glow blazing in the middle and split by a dark equatorial dust lane (known as the "hat brim") that gives this galaxy its name. As aperture increases into the 12- to 14-inch range, this becomes an object of ever-increasing wonder and delight, while the view in large observatory-class instruments actually surpasses photographs. Surprisingly, Herschel did not mention the prominent dust lane in his description although he surely must have seen it. The Sombrero is tilted to our line of sight by about 6° from being exactly edge-on, the dust lane passing S of the nuclear bulge – allowing us to see more of one half than the other, and giving it the surprising illusion of having depth or dimensionality. This amazing star city is estimated to contain nearly a *trillion* suns and is surrounded by more than 2,000 globular

Fig. 5.12. H I-43 (M104) is one of the brightest galaxies in the heavens, being visible in the smallest of telescopes even under bright-sky conditions. Better known as the Sombrero Galaxy, it was the first of several additions to Messier original list of 103 deep-sky objects that extended his catalog to its accepted number today of 110 entries. Courtesy of Mike Inglis.

clusters (as compared to fewer than 200 for our own Milky Way)! Surely, here is food for contemplation as you stare in wonderment at this marvel of the night (Fig. 5.12).

H I-39 (NGC 4697): 12 49 – 05 48, galaxy, 9.2, 7′ × 5′, = Caldwell 52. "*Very bright, large, little extended 45 degrees +/–, suddenly much brighter in the middle to a nucleus.*" This big elliptical system is typical of the multitude of galaxies strewn throughout Virgo. It is being included here because of its brightness and also because it is one of Sir Patrick Moore's Caldwell objects. Like most members of its class, it is an oval glow with a bright nucleus as seen in medium aperture scopes. Sweep for it exactly 5° due W of the telescopic multiple star θ Virginis, where you will find it positioned near a neat line of field stars.

H I-25 = H II-74 (NGC 4754): 12 52 + 11 19, galaxy, 10.6, 5′ × 3′. "*Bright, pretty large, round, pretty suddenly brighter in the middle, preceding of two.*" Here is a Herschel object that was entered in Class I at one time and then in Class II at another! Look for it 2°W and slightly N of the star Vindemiatrix (ε Virginis). In a 6-inch glass, this elliptical looks nearly circular and has a fairly prominent nucleus. Adding to the scene is another, dimmer Herschel object – the 10th-magnitude spiral H II-75 (NGC 4762), which he described as being "Pretty bright, very much extended 31 degrees, three bright stars south, following of two." It is 9′ × 2′ in apparent size and noticeably elongated with only a suggestion of a nucleus in small scopes. It lies just 11′ SW of H I-25/H II-74 and is encompassed by three field stars. Comparing it with its neighbor, which Herschel Class do you think better fits NGC 4754?

H I-70 (NGC 5634): 14 30 – 05 59, globular cluster, 9.6, 5′. "*Globular cluster of stars, very bright, considerably large, round, gradually brighter in the middle, well resolved, clearly consisting of stars, stars 19th magnitude, a star of 8th magnitude south following.*" Seemingly out of place in this galaxy-rich constellation but offering a nice change from observing faint fuzzies is this tiny, rather dim starball.

It lies 3°W of μ Virginis and just 30' E of 104 Virginis, which is located in the same wide eyepiece field. In small telescopes it looks like a slightly oval disk of subdued light, nicely placed close to a dim field star near its E edge. It takes instruments in the 12- to 14-inch aperture range and a steady night to see it as a little stellar beehive the way Herschel did – which brings up the question once again of why it was placed in Class I?

H I-126 (NGC 5746): 14 45 + 01 57, galaxy, 10.6, 8′ × 2′. *"Bright, large, very much extended 170 degrees, brighter in the middle to a bright nucleus."* This neat edge-on spiral is located just 30' W of the star 109 Virginis, making it easy to find. Its very thin cigar-shape is quite striking in 8-inch and larger scopes and a fairly obvious nucleus can be seen. Adding to the scene is another much dimmer Herschel object – the tiny 12th-magnitude spiral H II-538 (NGC 5740) just 18' SSW of H I-126 itself. The star and both galaxies can all be seen together within the same wide eyepiece field of view.

Chapter 6

Showpieces of Class IV

Planetary Nebulae

Listed below in alphabetical order by constellation are 29 of the most interesting objects in Herschel's Class IV. Following the Herschel designation itself is the corresponding *NGC* number in parentheses, its Right Ascension and Declination (for Epoch 2000.0), the object's actual type (which may differ from the Class Herschel assigned it to), its visual magnitude, angular size in minutes (′) or seconds (″) of arc, and Messier or Caldwell number plus popular name if any. Next is a translation of Sir William's shorthand description (in italicized quotes) taken from the *NGC* itself, followed by comments from the author. These include directions for finding each object by sweeping for it, just as Herschel himself originally did.

Andromeda

H IV-18 (NGC 7662): 23 26 + 42 33, planetary nebula, 8.5, 32″ × 28″, = Caldwell 22/Blue Snowball. "*Remarkable, a magnificent or otherwise interesting object, planetary or annular nebula, very bright, pretty small, round, blue.*" This gem is one of the brightest planetaries in the sky. Looking like a small blue dot in a 4-inch glass, it is quite vivid in an 8-inch glass. A 12-inch glass shows an annular or ring-like structure with a very faint 13th-magnitude star in the central void. Some observers claim this nucleus is variable since it is sometimes visible and sometimes not, but this has been attributed to changing seeing conditions. The nebula's hue has also been described as cobalt blue and bluish-green, but most see it as a pure blue ball – resulting in its popular name. It is easily located by sweeping along a line joining the stars ι and ο Andromedae, due N of the Great Square of Pegasus. You will find it in the same wide eyepiece field as the star 13 Andromedae, which lies less than 30′ to the NE. This celestial Snowball is some 5,600 light-years from us, positioned on the edge of the Milky Way where it courses through Andromeda (Fig. 6.1).

Aquarius

H IV-1 (NGC 7009): 21 04 – 11 22, planetary nebula, 8.3, 25″ × 17″, = Caldwell 55/Saturn Nebula. "*Remarkable, a magnificent or otherwise interesting object, planetary nebula, very bright, small, elliptic.*" Even brighter and more striking than H IV-18 is this magnificent, intensely greenish-blue cosmic egg. Easily spotted

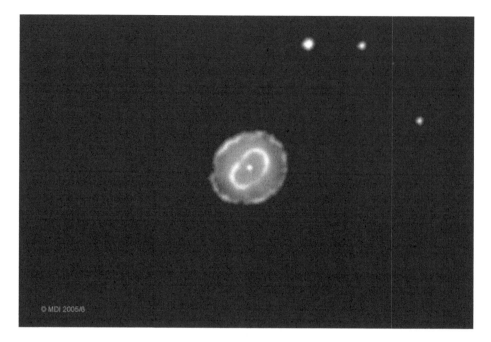

Fig. 6.1. H IV-18 (NGC 7662) is known as the Blue Snowball since it appears distinctly icy-blue to most observers (others seeing it as greenish-blue in color). Among the brightest planetaries in the sky, it is a fascinating sight in telescopes of all sizes. In large instruments like those Herschel used, it is truly an amazing sight! Courtesy of Mike Inglis.

just 2° due W of ν Aquarii, it is obvious even in a 3-inch glass and is a fascinating sight in 6-inch and larger telescopes. The ansae – or edge-on, ring-like extensions to the disk – need at least a 10-inch on a dark steady night to glimpse. The featureless disk itself has an eerie fluorescent radiance to it that is obvious in all apertures. There is actually a 12th-magnitude central star present, but it is all but drowned out by the brightness of the nebula itself. Note that Herschel mentioned neither the ansae nor the star. The Saturn Nebula is truly an amazing sight as seen in a 12- to 14-inch scope and the view in observatory-class instruments is quite beyond words. This was the first member of its exalted class to be discovered by Sir William – he must have been quite thrilled with what he had unexpectedly come across while sweeping the sky in Aquarius, and it surely encouraged him to search for more of its kind! This gem lies at a distance of 3,000 light-years from us (Fig. 6.2). (The dim globular cluster M72 and the little asterism M73 are located just a few degrees to the SW.)

Camelopardalis

H IV-53 (NGC 1501): 04 07 + 60 55, planetary nebula, 11.9, 55″ × 48″, Oyster Nebula. "*Planetary nebula, pretty bright, pretty small, very little extended, 1′ diameter.*" Deriving its name from its appearance on large-image-scale, short-exposure photographs, this faintish bluish-white disk looks like a pale gray

Fig. 6.2. One of the brightest and most spectacular planetaries in the heavens is H IV-1 (NGC 7009), better known as the Saturn Nebula. The first member of its class to be discovered by Sir William, it thrills observers today just as it must have done him upon finding it. Its eerie bluish-green color and elliptical shape are unmistakable in any telescope. Courtesy of Mike Inglis.

smudge in small scopes. In 10-inch and larger apertures, a dark center or smoke ring-like structure can be glimpsed and a 14th-magnitude central star has been reported by some observers. However, Herschel apparently did not see this dim nucleus in either of his 20-foot reflectors. Situated in a fairly isolated area, it can be found by carefully sweeping nearly 1 h of Right Ascension (or 15°) due W of the star β Camelopardalis. The small open cluster H VII-47 (NGC 1502 – see Chapter 9) lies about 2° to the N. Observers may find sweeping for the cluster first and then dropping S easier, since it is much more obvious than the nebula itself (Fig. 6.3).

Cassiopeia

H IV-52 (NGC 7635): 23 21 + 61 12, diffuse nebula, 7.0, 15′ × 8′, = Caldwell 11/Bubble Nebula. "*Very faint, a star of 8th magnitude involved a little excentric.*" Appearing as a faint, incomplete bubble on photographs, visually this object's brightest part is an arc only 3′ in size with an 8th-magnitude star embedded within it. Fortunately, it is situated just 36′SW of the beautiful open cluster M52, lying within the same wide eyepiece field. This makes locating it relatively easy. But actually glimpsing it is another matter! Visually, it is a difficult object to see in typical backyard telescopes, requiring a very dark moonless night and a well-dark-adapted eye. Note that Herschel himself even called it "very faint." It is best to look

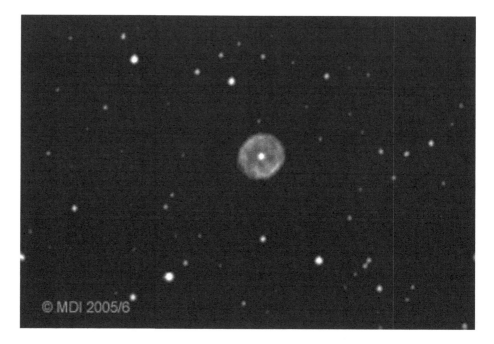

© MDI 2005/6

Fig. 6.3. H IV-53 (NGC 1501), better-known as the Oyster Nebula, is a large dim planetary that needs aperture to appreciate. It is also somewhat of a challenge to locate and requires careful sweeping to pick up. Courtesy of Mike Inglis.

for the 8th-magnitude star first and then use extreme averted vision to search for the ghostly nebulosity. The stars α and β Cassiopeiae point right at M52, a line through them extended its own length placing you right on top of it. This combo also lies just S of 4 Cassiopeiae. A final point – the author must confess puzzlement at just how this elusive object ever ended up rating a spot on the Caldwell list of showpieces!

Cepheus

H IV-76 (NGC 6946): 20 35 + 60 09, galaxy, 8.9, 11′ × 10′, = Caldwell 12. "*Very faint, very large, very gradually, then very suddenly brighter in the middle, partially resolved, some stars seen.*" Located about 2°SW of η Cephei at the Cygnus border is an object that has been described as a "mini M33" (the great spiral galaxy in Triangulum). While this may be the case photographically, visually it is fairly dim despite its listed magnitude (which is bright for a galaxy) due to its light being spread out over a large area, resulting in a low surface brightness. This glowing circular patch is somewhat difficult in a 6-inch and definitely needs a dark night to appreciate. It was misidentified as a planetary by Herschel – certainly something easily done from the appearance of many round galaxies in the eyepiece. But if it was partially resolved and some stars were seen as he noted, why was it logged as planetary nebulae? Adding to the scene is the open cluster H VI-42 (NGC 6939 – see Chapter 8) lying in the same wide eyepiece field just 38′ to the NW. Although these two wonders appear close together in the sky, they are actually vastly far apart in space, the galaxy being some 5,000 times the distance of the cluster (10,000,000 light-years compared to 2,000 light-years)!

H IV-74 (NGC 7023): 21 02 + 68 12, diffuse nebula, 6.8, 18′, = Caldwell 4/Iris Nebula. "*A star of 7th magnitude in an extremely faint, extremely large nebulosity.*" This object is one of the brightest reflection nebulae in the N sky, being obvious in a 5- or 6-inch glass even under conditions of moderate light pollution. In fact, at magnitude 6.8 (this being that of the involved star itself), it can be seen in large binoculars. The puzzle here is that Herschel described this nebula as "extremely faint." Its name stems from its appearance on photographs, where it does indeed look like a delicate bluish-purple celestial iris. This is one of the great objects omitted from the Herschel Club's original observing list. You can find it by sweeping 3°SW of the beautiful double star β Cephei, which marks the top right-hand corner of that constellation's distinctive house-shaped figure (Fig. 6.4).

H IV-58 (NGC 40): 00 13 + 72 32, planetary nebula, 10.2, 60″ × 40″, = Caldwell 2. "*Faint, very small, round, very suddenly much brighter in the middle, star of 12th magnitude south preceding.*" Here is one of the very few worthwhile planetaries to be found at high declinations. But being situated in an isolated region of the sky devoid of obvious stellar landmarks, it requires some careful star-hopping and sweeping to actually find. The author uses a line drawn from δ to ι Cephei prolonged its own length to arrive at its position. Slow sweeping about that point will bring it into view, situated between two faint field stars. While it can be seen in a 4-inch glass, at least twice that aperture is needed for a good view. Observers have variously reported its disk as being reddish, grayish, and greenish in hue – and all remarking that its 11.6th-magnitude central star is obvious as seen in 8-inch and larger scopes. Note that Herschel himself called this object "faint" and made no mention of an actual stellar nucleus as such. H IV-58 lies 3,000 light-years from us.

Fig. 6.4. H IV-74 (NGC 7023), the Iris Nebula, is not actually a planetary but rather a diffuse nebulosity. It surrounds a 7th-magnitude star, making it easy to spot even in the smallest of glasses. Courtesy of Mike Inglis.

Corvus

H IV-28.1 (NGC 4038): 12 02 – 18 52, galaxy, 10.7, 3′ × 2′, = Caldwell 60/Antennae/Ring-Tail Galaxy. H IV-28.2 (NGC 4039): 12 02–1852, galaxy, 12, 3′×2′, = Caldwell 61/Antennae/Ring-Tail Galaxy. *"Pretty bright, considerably large, round, very gradually brighter in the middle."/"Pretty faint, pretty large."* This pair of peculiar spirals is one of the easiest interacting/colliding galaxies in the sky to see in backyard telescopes. They appear as a single nebulous arc in scopes under 6-inch while in an 8-inch they look shrimp-shaped. Two faint extending filaments can be glimpsed in 12- to 14-inch apertures, although these were apparently not seen by Herschel himself. You will find this unusual combo about 3.5°SW of γ Corvi and 1°N of 31 Crateris. While the Caldwell listing recognizes both galaxies and assigns them separate numbers, the Herschel Club's original roster treats this as one object with a single designation – nor does it mention its popular name. We view this pair from across some 90,000,000 light-years of intergalactic space. Ironically, Corvus' only real planetary nebula, NGC 4361, ended up being placed in Class I as H I-65 (see Chapter 5)! (Fig. 6.5).

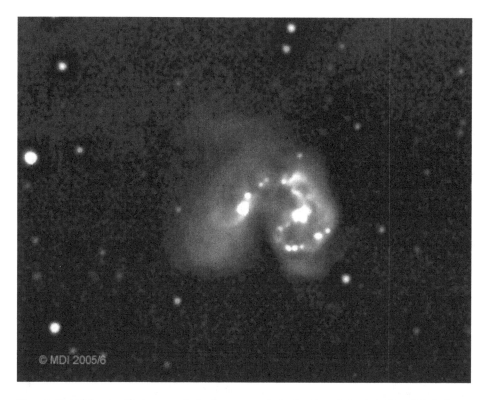

Fig. 6.5. While classified as a dual planetary nebula by Herschel, H IV-28.1/28.2 (NGC 4038/4039) is actually a pair of colliding/interacting galaxies known as the Antennae/Ring-Tail Galaxy. And as such, they rank among the easiest such objects to see in the sky using backyard telescopes. Their peculiar shrimp-shaped forms are a fascinating sight in medium to large instruments. Courtesy of Mike Inglis.

Cygnus

H IV-73 (NGC 6826): 19 45 + 50 31, planetary nebula, 8.9, 27″, = Caldwell 15/Blinking Planetary. *"Planetary nebula, bright, pretty large, round, a star of 11th magnitude in the middle."* In the August, 1963, issue of *Sky & Telescope* magazine, the author called attention to the amazing behavior of this object, coining the name Blinking Planetary. Here we find a bright, pale-blue disk with an obvious 10th-magnitude central star. Using direct vision to stare right at the star, the nebula itself becomes invisible. Changing to averted vision, the nebula snaps into view with an intensity that drowns out the star. Switching back and forth causes a thrilling apparent on-again, off-again blinking effect! This is one of the very few deep-sky objects that appear to *do* something as you watch them! The blinking can be seen in telescopes as small as 3- and 4-inch glasses, but the bigger the instrument the more spectacular it becomes. It is quite striking in 6- and 8-inch scopes. This effect was actually first noticed using a superb 13-inch Fitz-Clark refractor while conducting a visual sky survey for my *Sky & Telescope* series "The Finest Deep-Sky Objects" (which appeared in 1965 and 1966). In that historic instrument and others in its aperture range, H IV-73 is without question one of the most astonishing sights in the entire heavens! It is easily found just 45′E of the beautiful wide, golden double star 16 Cygni, lying in the same wide eyepiece field of view. A number of other planetaries exhibit a similar behavior, but none so strikingly as does this object. Herschel (and all the other classic observers) apparently never noticed this effect – which is really difficult to explain given its prominence. Is it possible that the nebula itself has physically evolved over the past 150–200 years enough to shift its primary emission lines into the retina-sensitive part of the spectrum, resulting in the blinking effect seen today? Also somewhat of a puzzle here is that, while the Herschel Club's observing list includes this object, it does not mention its name or its striking behavior (both of which are now very widely known to observers), its obvious central star, nor the striking double sitting near it. We watch this object's antics from a distance of some 3,300 light-years (Fig. 6.6).

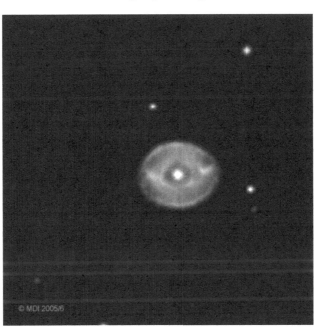

Fig. 6.6. H IV-73 (NGC 6826) is the amazing Blinking Planetary. Its seeming disappearance and reappearance results from switching back and forth between direct and averted vision, respectively. While this effect can be seen even in a very small telescope, it becomes ever-more spectacular as aperture increases. Courtesy of Mike Inglis.

H IV-72 (NGC 6888): 20 12 + 38 21, diffuse nebula, 8.8, 18′ × 13′, = Caldwell 27/Crescent Nebula. "*Faint, very large, very much extended, a double star attached.*" Also sometimes called the "Cygnus Bubble" from its appearance on photographs, this crescent-shaped object is much fainter than you might expect from its listed magnitude. This is due to its large apparent angular size, resulting in a low surface brightness. Note that Herschel himself called it faint and very much spread out. He also mentioned that a double star is attached to it. Because this object lies in a very rich region of the Milky Way, identifying the actual star-pair he was referring to is not easy. There is a 7th-magnitude star involved which provides the illumination for the nebula itself. This ghostly nebulosity is best viewed in 8-inch and larger apertures at low magnification on a dark night. It has been described as an oval ring as seen in observatory-class instruments. Look for it by sweeping about 2.5°SW of γ Cygni, the central star in the crossbar of the Northern Cross asterism. It also lies just 1°W of the well-known and enigmatic nova-like variable star P Cygni.

Delphinus

H IV-16 (NGC 6905): 20 22 + 20 07, planetary nebula, 11.9, 44″ × 38″, Blue Flash Nebula. "*Remarkable, very much so, planetary nebula, bright, pretty small, round, 4 small stars near.*" This small dim planetary can be seen in a 5- or 6-inch scope, appearing as an ill-defined bluish blob. But apertures twice that size are required to appreciate it, in which it appears distinctly blue with a ring-shaped structure. Its faint 14th-magnitude central star may also reveal itself, but Herschel does not mention it. All scopes show the presence of a number of surrounding field stars in this rich Milky Way region. Some observers actually refer to the nebula as lying in a course cluster, at a distance of over 4,000 light-years. It is easy to see why it is called "Blue" – but the reason for the "Flash" in its name eludes the author! A line drawn from δ through α Delphini points NW to its location near the intersection of the borders of Delphinus, Sagitta, and Vulpecula. Dropping 1°S of the star 29 Vulpeculae and then sweeping 3°W will bring you to it.

Draco

H IV-37 (NGC 6543): 17 59 + 66 38, planetary nebula, 8.8, 22″ × 16″, = Caldwell 6/Cat's Eye/Snail Nebula. "*Planetary nebula, very bright, pretty small, suddenly brighter in the middle to a very small nucleus.*" This showpiece can be readily seen in 3- and 4-inch glasses as a small but bright bluish-green ellipsoid. Telescopes 8- to 10-inches in size reveal hints of internal structure and its 10th-magnitude nuclear sun, which is not as easy to see as you might expect being immersed in the bright nebulosity. The star appears yellow due to a strong contrast with the nebula itself. Apertures in the 12- to 14-inch range at high magnification show a hint of intertwining rings and a dark round void right next to the central star. Herschel mentions the nucleus but not the hole. To find this emerald gem, draw a line from γ through ξ Draconis in the head of Draco and extend it about twice its length due N. Once there, sweep for it midway between δ and ζ Draconis. Lying 3,500 light-years from us, its dual names come from its striking appearance on observatory photographs (Fig. 6.7).

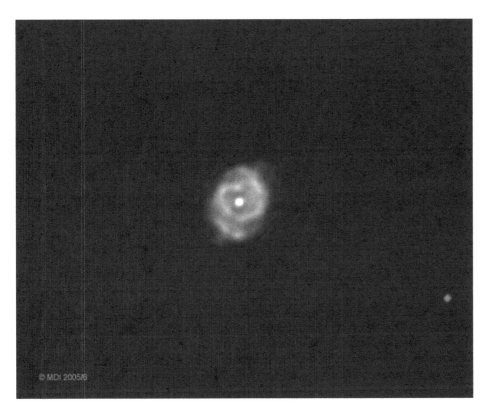

Fig. 6.7. H IV-37 (NGC 6543) is the famed Cat's Eye or Snail Nebula. This bluish-green cosmic egg is well named, especially as seen in photographs. While even a small glass reveals its disk and distinct hue, larger instruments show its central sun and fascinating internal structure. Being circumpolar, it is technically visible year-round – but like all deep-sky wonders, it is best-seen when high on the meridian. Courtesy of Mike Inglis.

Eridanus

H IV-26 (NGC 1535): 04 14 – 12 44, planetary nebula, 9.4, 20″ × 17″, Lassell's Most Extraordinary Object. "*Planetary nebula, very bright, small, round, pretty suddenly, then very suddenly brighter in the middle, resolvable (mottled, not resolved).*" The author gave this object its name from a remark made by the English observer William Lassell, who considered it to be the most extraordinary object of its kind he had ever seen in his homemade 24- and 48-inch metal-mirrored reflectors. (It has also been described as a "celestial jellyfish" floating in the ocean of space and as "another Neptune" from its photographic appearance.) Sweep for it about 30′N and 4°E of the star γ Eridani. It also lies just under 3° directly S of 39 Eridani. Appearing like a slightly out-of-focus star and somewhat difficult in a 3-inch scope at low power, a 6-inch shows its grayish-blue disk well and a 12-inch will reveal an 11.5-magnitude central star. Note that Herschel considered it resolvable. Interestingly, as of the time of writing, there are no published reports of its distance – this object seemingly having been totally ignored by astronomers despite Lassell's remark!

Gemini

H IV-45 (NGC 2392): 07 29 + 20 55, planetary nebula, 8.3, 20″, = Caldwell 39/Eskimo/Clown Face Nebula. *"Bright, small, round, a star of 9th magnitude in the middle, a star of 8th magnitude north following 100."* Now here is a friendly looking nebula if ever there was one (at least as seen on photographs)! This vivid blue planetary with a 10th-magnitude central star is visible in the smallest of glasses. And it is easily found, being situated just over 2°SE of the double star δ Geminorum. A finder, or very low-power/wide-angle eyepiece, shows two "stars" here – one is 63 Geminorum (itself a wide double star) and just to its S another fuzzy-looking one which is the nebula itself. (The real star, incidentally, makes a good target on which to focus before viewing the nebula itself.) Some observers consider this object to be a star enveloped in an atmosphere, which is an apt description as seen in small apertures. High magnification on telescopes in the 12- to 14-inch range reveals fascinating detail within this planetary, including overlapping bright rings and dark patches. This showpiece is some 3,000 light-years from us. One question here is why Herschel was not more excited by what he saw when he swept up this jewel. Surely, H IV-45 deserved at least his "remarkable" rating.

Hercules

H IV-50 (NGC 6229): 16 47 + 47 32, globular cluster, 9.4, 4′. *"Globular cluster of stars, very bright, large, round, disc and faint border, resolvable (mottled, not resolved)."* Here is one of the most fascinating cases of an object being misclassified by its discoverer (and with good reason, as we shall see) – and the misidentification being accepted and perpetuated by legions of observers until later research showed its true identity! You will find it located roughly midway between the stars 42 and 52 Herculis, due N of this constellation's Keystone asterism. At low power in small instruments this little ball does indeed look like a planetary nebula, forming a pretty triangle with two 6th-magnitude field stars. One account from the past describes it as beautifully grouped in a triangle and another calls it sea-green (the typical color of a planetary nebula) in a starry triangle. A 3-inch glass shows it neatly placed with its two stellar sentries, while scopes in the 6- to 8-inch range reveal it to be grainy (the precursor to actual resolution). But it takes a 10- to 12-inch and a steady night to transform this object into a sparkling starball. The mystery here is not that Herschel said it was resolvable (which he did for many nebulae) but that his description states right up front that it is a globular cluster – and yet he cataloged it as a planetary nebula. As we have previously seen, this and similar remarks may have been added later by his son Sir John or by Dreyer, the compiler of the *NGC*, in light of better observations. If so, it may well account for such seeming discrepancies found elsewhere in Herschel's various classes (especially his Class I). This object is small because it is so very remote for a globular, lying as it does some 90,000 light-years from us in our Galaxy's outer halo.

Hydra

H IV-27 (NGC 3242): 10 25 – 18 38, planetary nebula, 8.6, 40″ × 35″, = Caldwell 59/Jupiter's Ghost. *"Remarkable, planetary nebula, very bright, little extended 147 degrees, 45" diameter, blue."* Also known as the Eye Nebula and the CBS Nebula from its appearance on photographs (and as seen in large observatory telescopes), visually this big bright planetary displays a pale blue disk roughly the same apparent size as the planet Jupiter. It is among the easiest of its class to see, being obvious even in a 2.4-inch (60-mm) glass. Its 11th-magnitude central sun and lots of internal structure (including a double-ring that some observers report sparkles) become visible in 10-inch and larger apertures. Surprisingly, none of this is mentioned by Herschel in his description even though his telescopes certainly had the capability of showing these features. Regarding the nucleus itself, the noted observer William H. Pickering reported in 1892 that the nebula had a bright star in its center but that by 1917 it had disappeared! It is actually still there and – as is the case for the nuclei of many planetaries – it is suspected of being variable in brightness, magnitudes ranging from 10.3 to 11.4 being reported. Perhaps when Sir William discovered this object, the star had faded and disappeared from sight as it did in Pickering's case. It should be pointed out here, however, that observing the central stars of planetaries is often difficult and uncertain due to interference from the bright encompassing nebulosity. This is one of the best objects of its class that Messier missed, being noticeably brighter than the famed Ring Nebula (M57) in Lyra. It is easily found less than 2°S of the star μ Hydrae, some 40° due S of the Sickle asterism in Leo, and is 3,300 light-years away.

Monoceros

H IV-2 (NGC 2261): 06 39 + 08 44, diffuse nebula, 10.0, 2′ × 1′, = Caldwell 46/Hubble's Variable Nebula. *"Bright, very much extended 330°, nucleus cometic equals a star of 11th magnitude."* This little comet- or fan-shaped nebula with the dim erratic variable star R Monocerotis embedded at its S tip lies 1°S of the star 15 Monocerotis – itself involved in the big bright Christmas Tree Cluster (H VIII-5/NGC 2264 – see Chapter 10). Its brightness, size, and shape all change with variations in R itself, something discovered in 1916 by the famed astronomer Edwin Hubble. While the nebula is a fascinating sight in 8-inch and larger apertures, the variable star is often difficult to pick out from the surrounding nebulosity. The use of high magnifications is recommended whatever be the size of the instrument being employed. In large backyard scopes a bluish tint to the nebula may be noted. This strange object lies some 2,600 light-years from us, immersed in the star-stream of the Winter Milky Way (Fig. 6.8).

Ophiuchus

H IV-11 (NGC 6369): 17 29 – 23 46, planetary nebula, 11.5, 30″, Little Ghost Nebula. *"Remarkable, very much so, annular nebula, pretty bright, small, round."* As one observer puts it, this ghostly little ring-shaped nebula is easy to "scare up"! Look

Fig. 6.8. While cataloged as a planetary nebula by Herschel, H IV-2 (NGC 2261) is actually a diffuse nebulosity – and a very famous one at that! Known as Hubble's Variable Nebula, its strange fan-shaped form can be glimpsed in medium apertures and at times even its imbedded illuminating star. Courtesy of Mike Inglis.

for it just 30′NW of the star 51 Ophiuchi, which is itself 2°NE of θ Ophiuchi. Generally regarded as a dim miniature of the Ring Nebula (M57) in Lyra, it is not only much smaller than its famous counterpart but much fainter. It can be seen in a 4-inch glass, and appears as a perfect little smoke ring at high power in 6- to 8-inch telescopes. A pale blue or green hue is also generally noticeable. Its extremely faint 15th-magnitude central star was not seen by Herschel and lies beyond the reach of all but the very largest amateur instruments. The Little Ghost shimmers at us from across a gulf of some 3,800 light-years.

Orion

H IV-34 (NGC 2022): 05 42 + 09 05, planetary nebula, 12.0, 18″. "*Planetary nebula, pretty bright, very small, very little extended.*" Here is a faint ring-shaped planetary located near the head of Orion for viewing with medium to large backyard telescopes. Look for it 2°SE of the attractive double star λ Orionis. Two dim stars will be seen in a finder or wide-field eyepiece about 30′ to the NW of the nebula itself, seeming to point right at it. Even so, careful sweeping on a dark, transparent night is needed to pick it up. A 10-inch at high power will reveal this object's ring-like structure using averted vision – something Herschel did not mention seeing himself. Some observers have reported a bluish tint to the nebula, but it looks pale white or gray to the author. Its elusive central star is fainter than 14th-magnitude and is never mentioned in descriptions of this object. (It was not seen by Herschel either.) Despite Orion's riches, this is the only planetary nebula in this constellation bright enough to be seen in typical backyard instruments. The distance of this challenging object is 7,000 light-years.

Puppis

H IV-39 (NGC 2438): 07 42 – 14 44, planetary nebula, 11.0, 66″, In Messier 46.
"Planetary nebula, pretty bright, pretty small, very little extended, resolvable (mottled, not resolved), 3.75′ diameter." Here is a great example of that wonderful class of deep-sky objects in which two targets lie within the same field of view. This little dim planetary is projected against the lovely open cluster, M46. You will find them just 1°W of the close pair of stars 2 and 4 Puppus. The nebula can be glimpsed in a 4- or 5-inch glass near the N edge of the stellar commune, while an 8 inch shows it as tiny gray smoke-ring with a dim background star shining through its central hole. While the nebula looks like it lies within the cluster itself, this is only a chance alignment! The latter is 5,400 light-years from us and the former much closer at 3,000 light-years. Although their actual distances were not known in Herschel's time, he brilliantly deduced *solely from their appearance in the eyepiece* that the planetary has "No connexion with the cluster, which is free of nebulosity"! Surprisingly, however, he mentioned neither the ring-like shape of this planetary nor its apparent central star despite their being readily visible in modest amateur instruments. On the other hand, note his remark about the nebula being "resolvable"(Fig. 6.9).

Fig. 6.9. H IV-39 (NGC 2438) is a tiny dim, ring-shaped planetary suspended against the stars of the rich open cluster M46. Telescopes in the medium to large range are needed to really appreciate this celestial smoke-ring. Courtesy of Mike Inglis.

H IV-64 (NGC 2440): 07 42 – 18 13, planetary nebula, 10.5, 50″ × 20″. "*Planetary nebula, considerably bright, not very well defined.*" Appearing star-like in a 3-inch glass at low power, a 6-inch shows this object as a round patch of bluish or turquoise light about 20″ in diameter. Larger apertures at high power reveal a faint irregular or rectangular glow surrounding the brighter center. There is also a very dim 14th-magnitude central star, which is rarely mentioned by observers. Visual magnitude estimates for this planetary are highly discordant, ranging from as bright as 9.3 to as faint as 11.5 in. magnitude. Its distance is 3,500 light-years.

Sagittarius

H IV-41 (NGC 6514): 18 03 – 23 02, diffuse nebula, 6.3, 28′, = Messier 20/Trifid Nebula. "*Remarkable, a magnificent or otherwise interesting object, very bright, very large, trifid, a double star involved.*" Now here is a real puzzle. This object was certainly known to Herschel as M20, but he gave it four separate designations of his own – that of a planetary as discussed here and three as a very large nebula (see Chapter 7). His description provides no clue as to what part of the nebula he had actually assigned his numbers to – in fact, it simply describes the Trifid as a whole. (His original paper in the *Philosophical Transactions* of the Royal Society does likely make these clear.) So, as observers we are left to examine this big dim cloud and see if we can find anything that has the appearance of a planetary nebula. The double star Herschel refers to is known as HN 40. It is actually a triple system, consisting of 7th-, 8th-, and 10th-magnitude components just 11″ and 5″ apart. (Apparently the dim star at 5″ was not resolved in his reflectors.) The Trifid is positioned 1.5°N of the much brighter Lagoon Nebula (M8), which is visible to the unaided eye and in binoculars on a dark night above the spout of this constellation's Teapot asterism. We view both nebulae from a distance of 5,500 light-years.

H IV-51 (NGC 6818): 19 44 – 14 09, planetary nebula, 9.9, 22″ × 15″, Little Gem Nebula. "*Planetary nebula, bright, very small, round.*" This is a small but strongly tinted planetary located 2°N of the pair of stars 54 and 55 Sagittarii. It appears stellar in 2- to 4-inch scopes at low power but reveals a tiny disk when magnified. As aperture increases, this object does look like some exquisite piece of cosmic jewelry. Its color has been variously described as bluish, greenish, and turquoise. Most telescopes show it as an intense, uniform blue disk but large ones reveal hints of an internal ring-like structure. The visual brightness (actually dimness!) of its central star is highly uncertain, values ranging from 13th- to 15th-magnitude being reported. Adding to the interest here, in the same wide eyepiece field 45′SSE of the Little Gem, is a very famous but challenging object known as Barnard's Dwarf Galaxy (NGC 6822). It was missed by Herschel (see Chapter 12) and with good reason – this is not an easy object! It is an extremely faint 16′ × 14′ ghostly smudge of light. Although discovered with a 5-inch glass by the eagle-eyed Barnard, most observers find that it takes at least an 8-inch, a wide field of view, and a very dark, transparent night to glimpse it. The planetary itself is 5,000 light-years from us.

Taurus

H IV-69 (NGC 1514): 04 09 + 30 47, planetary nebula, 10.9, 2′. *"Star of 9th magnitude in nebula 3' in diameter."* This unusual object brings to light another case of brilliant deduction by William Herschel reminiscent of that of the nebula-cluster combo H IV-39 (NGC 2438) and M46. Let us begin by first discussing how to find it. This planetary is located 2°N of ψ Tauri. A line drawn from the star o through ζ Persei points to its position, sitting virtually on the Taurus–Perseus border. Sweeping here, a 4-inch glass shows a faint but obvious nebulous halo surrounding a 9th-magnitude star, bracketed by two field stars. In an 8-inch it becomes an intriguing sight, and it has been likened to a crystal ball as seen in a 10-inch Although not a spectacular sight in itself, this object has great historical significance. In order to appreciate this, we now quote more fully from Sir William's original complete description of H IV-69 rather than the shorthand version above taken from the *NGC*: "A most singular phenomenon! A star of about 8th-magnitude with a faint luminous atmosphere, of circular form, and about 3 minutes in diameter. The star is in the centre, and the atmosphere is so faint and delicate and equal throughout that there can be no surmise of it consisting of stars; nor can there be a doubt of the evident connexion between the atmosphere and the star." Thus he recognized for the first time the existence of "a shining fluid of a nature totally unknown to us" as he described it. Until this observation, all nebulae were thought to be simply unresolved masses of stars. This amazing deduction – *based solely on the appearance of this object in the eyepiece of his telescope* – showed nebulae to be gaseous long before the spectroscope actually proved it (Fig. 6.10)!

Fig. 6.10. H IV-69 (NGC 1514) is historically important, for the telescopic appearance of this planetary led Herschel to conclude that the nebulosity surrounding its central sun was physically real and not just an unresolved mass of stars – which all nebulosities were believed to be at the time. Courtesy of Mike Inglis.

Ursa Major

H IV-79 (NGC 3034): 09 56 + 69 41, galaxy, 8.4, 11′ × 5′, = Messier 82. "*Very bright, very large, very much extended (a ray).*" As in the case of H IV-41 for the Trifid nebula (M20), this peculiar cigar-shaped galaxy is a well-known Messier object to which Herschel strangely assigned a number in his class of planetary nebulae. His notation clearly described the galaxy itself, so what does his planetary designation refer to? M82 and its spiral companion M81 together form one of the most striking and easily-observed galaxy pairs in the sky (actually being visible in binoculars on a dark night!), both objects fitting into the same wide eyepiece field 30′ apart. A line drawn from γ through α Ursae Majoris in the bowl of the Big Dipper extended its own length brings you to their approximate position. Slow sweeping of this area readily picks them up. They serenely sail the ocean of intergalactic space at a distance of some 11,000,000 light-years.

H IV-61 (NGC 3992): 11 58 + 53 23, galaxy, 9.8, 8′ × 5′, = Messier 109. "*Considerably bright, very large, pretty much extended, suddenly brighter in the middle, bright resolvable (mottled, not resolved) nucleus.*" This neat spiral was not assigned a Messier number until long after Herschel observed it, so he understandably thought it was an original discovery. But (at least in light of modern knowledge) this object looks much more like a galaxy than a typical planetary nebula. It is an easy catch in 4- to 6-inch telescopes, and is noticeably elongated. Telescopes twice that aperture show a brighter center and at high magnifications hints of internal structure, which has the form of the Greek letter θ on photographs. It is easily spied in the same wide eyepiece field as γ Ursae Majoris, 40′ to its SE (Fig. 6.11).

Fig. 6.11. H IV-61 (M109) is a fairly bright, egg-shaped spiral galaxy that is a snap to find since it lies in the same wide field of view as γ Ursae Majoris – one of the stars marking the Big Dipper's bowl. Courtesy of Mike Inglis.

@ MDI 2005/6

Fig. 6.12. H IV-8/9 (NGC 4567/4568) was included in Herschel's class of planetaries but this dual object is actually a close pair of spirals known as the Siamese Twins. Like H IV-28.1/28.2 in Corvus, this is one of the few interacting galaxies that can be observed in small telescopes. But to see them as clearly separate objects (yet still joined at their tips – thus this pair's name) requires at least a medium-sized glass. Courtesy of Mike Inglis.

Virgo

H IV-8 (NGC 4567): 12 36 + 11 15, galaxy, 11.3, 3′ × 2′, Siamese Twins (with H IV-9).

H IV-9 (NGC 4568): 12 37 + 11 14, galaxy, 10.8, 5′ × 2′, Siamese Twins (with H IV-8). *"Very faint, large, north preceding of a double nebula."/"Very faint, large, south following of a double nebula, position [angle of pair] 160 degrees +/–."* This pair of relatively dim spirals is in actual contact with other, and their photographic appearance certainly justifies the name given to them. Visually, large backyard telescopes show separate galaxies touching each other, but in small instruments (those 8-inches and under) they appear merged into a distorted fuzzy blob of light. Look for the Twins 2°NW of the star-pair ρ and 27 Virginis, and 1°S of the galaxy M58. This places them in the thick of the Virgo Galaxy Cluster, so careful star-hopping and sweeping is needed to identify them. As one of the few interacting galaxies within reach of small telescopes, they are certainly worth searching out (Fig. 6.12).

Chapter 7

Showpieces of Class V

Very Large Nebulae

Listed below in alphabetical order by constellation are 26 of the most interesting objects in Herschel's Class V. Following the Herschel designation itself is the corresponding *NGC* number in parentheses, its Right Ascension and Declination (for Epoch 2000.0), the object's actual type (which may differ from the Class Herschel assigned it to), its visual magnitude, angular size in minutes (′) or seconds (″) of arc, and Messier or Caldwell number plus popular name if any. Next is a translation of Sir William's shorthand description (in italicized quotes) taken from the *NGC* itself, followed by comments from the author. These include directions for finding each object by sweeping for it, just as Herschel himself originally did.

Andromeda

H V-18 (NGC 205): 00 40 + 41 41, galaxy, 8.0, 17′ × 10′, = Messier 110/Companion to M31. "*Very bright, very large, much extended 165 degrees, very gradually very much brighter in the middle.*" This elliptical system had not been attributed to Messier in Herschel's time, so he accordingly logged it as a new discovery (actually made by his sister Caroline). As one of the two close-in companions to the great Andromeda Galaxy (M31), it is readily located within the same field of view 36′ NW of M31's bright nuclear hub. Both much larger and dimmer in appearance than Andromeda's other elliptical companion, M32, it can be glimpsed using averted vision in a 2- to 3-inch glass as a pale oval glow lying outside the main galaxy's boundary. In larger scopes, it looks pearly-white and some observers have reported actually seeing it sparkle in big backyard instruments. We view this spectacular trio of "island universes" from across a gulf of 2,500,000 light-years. M110 is the most recent addition to Messier's original list, having been added in 1967, and will likely be the last ever made (Fig. 7.1).

H V-19 (NGC 891): 02 23 + 42 42, galaxy, 9.9, 14′ × 3′, = Caldwell 23. "*Remarkable, bright, very large, very much extended 22 degrees.*" Due to a rich starry foreground creating a striking 3D effect in which the galaxy seems to be floating among the stars, this is the most picturesque edge-on spiral in the entire sky as seen on photographs. But visually, it is not an easy object to find or see in small telescopes. It lies roughly midway between the magnificent double star Almach (γ Andromedae) and the pretty open cluster M34 in neighboring Perseus. Working from γ itself,

Fig. 7.1. H V-18 (M110) is the larger and fainter of the two close-in companions to the Andromeda Galaxy, the other one being M32. Discovered by Caroline Herschel, it was later found that Messier had actually seen it but not cataloged it. Courtesy of Mike Inglis.

sweep 3° due E. While it has been seen in 4-inch and even smaller scopes, at least a 6-inch on a very dark night is needed to glimpse it under typical light-polluted skies. In 10-inch and larger apertures, it appears as a ghostly cigar-shaped glow amid a curtain of faint stars, and the dark equatorial band so strikingly shown on photos can be made out. Note that Herschel apparently did not see this band, which runs the entire length of the galaxy. The actual visual magnitude of this phantom of the night seems very uncertain, values ranging from as bright as 9th- to as faint as 12th-magnitude being given. (It definitely seems *much* fainter in the eyepiece than the 9.9-magnitude listed above.) There is also considerable uncertainty about its distance, values ranging from 13,000,000 to 43,000,000 light-years appearing in the literature (Fig. 7.2).

Antlia

H V-50 (NGC 2997): 09 45 – 31 11, galaxy, 10.6, 8′ × 6′. *"Remarkable, very faint, very large, very gradually then very suddenly brighter in the middle to a nucleus of 4″, 19.5 seconds diameter."* This relatively dim spiral is one of William Herschel's southernmost discoveries. Despite the fact that he called this object "very faint," he still considered it to be "remarkable." This denizen of extragalactic space definitely requires aperture to be seen well – plus a dark transparent night and an unobstructed southern view. And like all celestial objects, it should be viewed

Fig. 7.2. Although H V-19 (NGC 891) is one of the most photogenic galaxies in the sky, visually it is not an easy object in small instruments. Seeing this edge-on spiral's dark equatorial band and rich curtain of foreground stars that make it so striking in photographs requires a really large backyard telescope. Courtesy of Mike Inglis.

when on or near the celestial meridian and, therefore, at its highest altitude in the sky. It can be found by sweeping 3°S of the star θ Antliae, or 2.5°NE of the pair of stars ζ-1 and ζ-2 Antliae, floating all alone in the sky. A puzzle here is what the "19.5 seconds diameter" mentioned in Herschel's description refers to. Note that the "seconds" are seconds of Right Ascension (or time) rather than seconds of arc (or angle).

Camelopardalis

H V-44 (NGC 2403): 07 37 + 65 36, galaxy, 8.4, 18′ × 11′, = Caldwell 7. "*Remarkable, very much so, considerably bright, extremely large, very much extended, very gradually much brighter in the middle to a nucleus.*" This striking spiral is among the best non-Messier galaxies in the sky. But due to its isolated position high in the N-circumpolar region, it is also frequently overlooked by observers. Drawing a line from the star-pair 26 and θ Ursae Majoris through o Ursae Majoris and extending it about two-thirds of its length leads you to its location. Sweeping about 1°W of the star 51 Camelopardalis will bring it into view. Actually visible in binoculars as a hazy spot of light, its elliptical glow is a lovely sight even in a 3- or 4-inch glass. In 8-inch and larger apertures a mottled texture can be seen surrounding its nucleus – something Herschel surprisingly did not comment about. It has been described as an "ocean of turbulence" when viewed at high powers. Overall, this showpiece resembles a smaller version of the Triangulum/Pinwheel Galaxy, M33. It lies 12,000,000 light-years from us (Fig. 7.3).

Fig. 7.3. H V-44 (NGC 2403) is a big bright spiral that was overlooked by Messier and left to William Herschel to discover. Its high northerly declination and relative isolation results in an undeserved obscurity, for it is one of the best of its class in the sky. Courtesy of Mike Inglis.

Canes Venatici

H V-41 (NGC 4244): 12 18 + 37 49, galaxy, 10.2, 16′ × 2′, = Caldwell 26. "*Pretty bright, very large, extremely elongated 43 degrees, very gradually brighter in the middle.*" This spiral is one of the flattest galaxies known, appearing as an extremely narrow streak or sliver in 6-inch and larger telescopes. This thin celestial ray is easily picked up by sweeping about 2°SW of 6 Canum Venaticorum. Both this object and the galaxy itself lie due W of the very pretty double star α Canum Venaticorum, better known as Cor Caroli (Fig. 7.4).

H V-43 (NGC 4258): 12 19 + 47 18, galaxy, 8.3, 18′ × 8′, = Messier 106. "*Very bright, very large, very much extended 0 degrees, suddenly brighter in the middle to a bright nucleus.*" Here is another later addition to the Messier list that Herschel discovered independently. This tightly-wound spiral lies about 1.5° due S of the star 3 Canum Venaticorum. A line extended from neighboring α through γ Ursae Majoris in the Big Dipper's bowl points right at it. This object has a high surface brightness despite its large size and is easily spotted in the smallest of telescopes. A 6-inch glass shows it as a large, luminous oval or pear-shaped glow with an obvious nucleus. Comments like "big and bold," "whopping apparent size" and "a grand galaxy to contemplate" typify the excitement this wonder elicits from observers. It lies at a distance of some 33,000,000 light-years.

H V-42 (NGC 4631): 12 42 + 32 32, galaxy, 9.3, 15′ × 3′, = Caldwell 32/Humpback Whale Galaxy. "*Remarkable, very bright, very large, extremely extended 70 degrees*

Fig. 7.4. H V-41 (NGC 4244) is among the flattest galaxies known, appearing as a narrow ray in medium-aperture instruments. It is another edge-on spiral like H V-19 but without a prominent dark equatorial band. Courtesy of Mike Inglis.

+/–, brighter in the middle to a nucleus, a star of 12th-magnitude attached to north." Messier would surely have regretted missing this grand object, for it is among the best galaxies in the sky and one of the largest edge-on spirals known. Indeed, it is considered by the Herschel Club to be "one of the prized Herschel objects." Sweep for it midway between the stars Cor Caroli (α Canum Venaticorum) and γ Comae Berenices – the brightest star in the neighboring naked-eye Coma Star Cluster. It also lies about 30′ NW of H I-176/177 (NGC 4656/4657), better-known as the Hockey Stick Galaxy (see Chapter 5). Both objects neatly fit within the same wide eyepiece field, each looking very long and ray-like. This object's name comes from its unusual lumpy shape, which is evident even in a 4-inch glass. Larger apertures show that the Humpback has a definite mottled look to it and that it is missing the typical edge-on dust lane down its long axis. Such scopes may also disclose a tiny companion galaxy lying closely above (N of) its mid-section. This is the 12th-magnitude elliptical NGC 4627, which Herschel apparently did not see. That is unless the 12th-magnitude "star" he mentioned refers to the galaxy. But this seems unlikely since there *is* such a star seemingly attached (as he stated) to the galaxy's edge, between it and the little elliptical itself. We view this celestial whale swimming the ocean of space from across a distance of 25,000,000 light-years (Fig. 7.5).

Fig. 7.5. H V-42 (NGC 4631) is popularly known as the Humpback Whale Galaxy from its unusual lumpy shape. An easy catch even in small telescopes, it is one of the brightest and best galaxies missed by Messier – a real showpiece of its class! Courtesy of Mike Inglis.

Cetus

H V-25 (NGC 246): 00 47 – 11 53, planetary nebula, 8.5, 4′, = Caldwell 56. "*Very faint, large, 4 stars in diffuse nebula.*" This unusual ephemeral-looking object could easily pass for a diffuse nebula, but it is actually a planetary and so technically belongs in Class IV. At 4′ in diameter, it is big for a member of its class and its light is so spread out that it has an extremely low surface brightness. Yet, it has been glimpsed in a 4-inch glass. In 10-inch and larger telescopes, what appears to be a complete ring can be seen, embedded in and surrounding which are several faint stars. Photographs also show several circular dark areas within the nebulosity. It is located 6°N of β Ceti, where it forms an equilateral triangle with φ-1 and φ-2 Ceti, the nebula lying at the S tip of the triangle. In the same wide eyepiece field 15′ to the NNE is the tiny, dim 12th-magnitude spiral H II-472 (NGC 255) (Fig. 7.6).

H V-20 (NGC 247): 00 47 – 20 46, galaxy, 8.9, 20′ × 7′, = Caldwell 62. "*Faint, extremely large, very much extended 172 degrees.*" This huge spiral appears as a very dim elongated glow orientated N–S in small telescopes. It is an outlying member of the small galaxy cluster centered in neighboring Sculptor, which includes the magnificent showpiece H V-1 (NGC 253), better-known as the Sculptor Galaxy (see the description below under that constellation). H V-20 lies at roughly the same distance from us as it and the other galaxies in the cluster. Sweep for it 3°S and slightly E of β Ceti, just above a starry triangle as seen in a finder. Some sources list this object as faint as nearly 11th-magnitude (which it certainly is not), its apparent dimness stemming from its large size and resulting low surface brightness (Fig. 7.7).

Fig. 7.6. H V-25 (NGC 246) is a big pale planetary nebula requiring a dark night and lots of aperture to be seen well. Although seemingly bright at magnitude 8.5, its light is spread out over such a large area of the sky that it appears much fainter visually than would be expected from its listed brightness. Courtesy of Mike Inglis.

Fig. 7.7. H V-20 (NGC 247) is a big edge-on spiral galaxy that, despite its relatively bright listed magnitude, appears rather dim in the eyepiece since its light is spread out over a large angular area resulting in a low surface brightness. It is also overshadowed by the huge bright neighboring showpiece, H V-I (NGC 253), the Sculptor Galaxy. Courtesy of Mike Inglis.

Coma Berenices

H V-24 (NGC 4565): 12 36 + 25 29, galaxy, 9.6, 16′ × 3′, = Caldwell 38. *"Bright, extremely large, extremely extended 135 degrees, very suddenly brighter in the middle to a nucleus equal to a star of 10th to 11th magnitude."* Wow – now here is a sight never to be forgotten! It is easily swept up slightly less than 2° due E of the wide double 17 Comae Berenices in the large naked-eye Coma Star Cluster. This grand object is a nearly perfectly edge-on spiral with a bulbous center and prominent dark dust lane splitting it into two lengthwise. It appears as a fairly dim narrow streak or needle of light in a 4- to 5-inch glass, which can show the dust lane with careful use of averted vision. In 8- to 10-inch apertures this galaxy is quite striking, while in 12- to 14-inch scopes it is truly a wondrous sight, looking much like its photographs! As the very first entry in his Class V, this marvelous object must surely have given Sir William a thrill upon seeing it. Note however, that he neither mentioned the dust lane nor considered it remarkable! We marvel at this extragalactic jewel from across a distance of some 20,000,000 light-years (Fig. 7.8).

Cygnus

H V-15 (NGC 6960): 20 46 + 30 43, supernova remnant, 7.9, 70′ × 6′, = Caldwell 34/Veil/Filamentary Nebula. *"Remarkable, very much so, pretty bright, considerably large, extremely irregular figure, Kappa Cygni involved."*/**H V-14 (NGC 6992/5): 20 56 + 31 43, supernova remnant, 7.5, 60′ × 8′, = Caldwell 33/Veil/Filamentary Nebula.** *"Remarkable, very much so, extremely faint, extremely large, extremely extended, extremely irregular figure, bifurcated."* This dual entry consists of two large arcs of nebulosity that are part of a huge bubble or loop (giving rise to another name sometimes used for them – the Great Cygnus Loop) that likely resulted from a supernova explosion that is estimated to have occurred as

Fig. 7.8. H V-24 (NGC 4565) is a large, spectacular edge-on spiral galaxy split its entire length by a dark equatorial dust lane and displaying an obvious nuclear bulge. Impressive even in small scopes, it is best enjoyed on dark transparent nights. Courtesy of Mike Inglis.

recently as 5,000, or as long ago as 150,000, years! It lies at the W tip of a squat triangle formed with the stars ζ and ε Cygni. Another way to pin it down is to look for its SW half H V-15 first, since it is centered on the unequal double star 52 Cygni (which Herschel referred to as κ). This section appears as a nebulous ray extending N–S of the star itself. The other half, H V-14, appears somewhat bigger but is harder to identify. The arcs themselves are more than 1° in length and are separated by about 2.5°. Both halves can be seen together in big binoculars and rich-field telescopes (RFTs) on a dark night. Individually, these ghostly wraith-like filaments are somewhat difficult and indistinct in a standard 6-inch scope. However, the use of a nebula filter transforms their appearance, making them quite obvious. The largest amateur instruments so equipped are claimed to show as much detail as do photographs. If this wonder is indeed the result of a long-ago supernova, it ranks along with the Crab Nebula (M1) in Taurus as one of the brightest such remnants in the entire sky visible in backyard telescopes. Very surprisingly, however, it is not included in the original Herschel Club listing. Also known as the Lacework Nebula and the Cirrus Nebula, this celestial soap bubble lies 1,500 light-years from us (Fig. 7.9).

H V-37 (NGC 7000): 20 59 + 44 20, diffuse nebula, 5.0, 100′ × 60′, = Caldwell 20/North America Nebula. "*Faint, extremely extremely large, diffused nebulosity.*" Just how Herschel managed to find this huge nebulosity given the very limited fields of view of his big reflectors is something of a mystery to the author. This gossamer glow lies about 3° due E of radiant, bluish Deneb (α Cygni) and measures nearly 2° in greatest extent (or four times the diameter of the full Moon!). Its presence and shape is best made out in binoculars or rich-field telescopes (RFTs) in the 3- to 6-inch aperture range. Bigger instruments typically loose the North-American-continent outline due to their smaller fields of view and the reduced contrast stemming from their larger image scales. This object needs a very dark transparent night to be seen in its full glory. It is estimated to lie 1,600 light-years from us (Fig. 7.10).

Fig. 7.9. HV-14 (NGC 6992/5) is the NE half of the huge supernova bubble known as the Veil or Filamentary Nebula. Its other half is H V-15 (NGC 6960), which lies 2.5° to the SW. Both of these ghostly nebulous arcs can be seen together in big binoculars and wide-field telescopes. Only a portion of H V-14 is shown in this detailed image. Courtesy of Mike Inglis.

Fig. 7.10. H V-37 (NGC 7000), the huge North America Nebula, requires a dark night and a very wide field of view to be seen. Herschel certainly had dark nights but how he ever managed to find this object given the small fields of view of his big reflectors is a mystery. Photographed using a 300-mm *f*/2.8 camera. Courtesy of Steve Peters.

Draco

V-51 (NGC 4236): 12 17 + 69 28, galaxy, 9.6, 20′ × 8′, = Caldwell 3. *"Very faint, considerably large, much extended 160 degrees +/−, very gradually brighter in the middle."* This big edge-on spiral has one of the largest apparent sizes known among galaxies. It is located well above the Big Dipper's bowl, slightly less than 2° due W of the little knot of stars κ, 4 and 6 Draconis. Despite being rated at magnitude 9.6 (some sources make it a full magnitude fainter), it appears quite dim due to its great size and resulting low surface brightness. While it can be glimpsed in a 6-inch glass, this ghostly sliver of light really needs aperture to appreciate.

Fornax

H V-48 (NGC 1097): 02 46 − 30 17, galaxy, 9.2, 9′ × 7′, = Caldwell 67. *"Very bright, large, very much extended 151 degrees, very bright in the middle to a nucleus."* Here we find a bright barred spiral whose luminous bar gives it a more elongated appearance in the telescope than the angular dimensions given here would suggest. As is typical for such objects, there is a faint spiral arm extending from each end of the bar. Although Herschel did not mention seeing them (which is not at all surprising considering how far S this object lies), large backyard instruments suggest their presence on a dark night using averted vision. It can be found by sweeping just 2°N of β Fornacis, floating alone in relative isolation.

Leo

H V-8 (NGC 3628): 11 20 + 13 36, galaxy, 9.5, 15′ × 4′. *"Pretty bright, very large, very much extended 102 degrees."* This big lovely edge-on spiral lies in the same wide eyepiece field as the galaxy pair M65 and M66, just 30′ to their N. Together, these three objects are popularly known as the Leo Triplet of spirals. All three can be seen in a 3-inch glass at low power, and they are a striking sight in 6-inch and larger scopes. Telescopes 10-inch and larger show H V-8's long thin glow bisected by a dark dust lane. Yet, Herschel mentioned neither this band, nor the presence of the two Messier galaxies in his description since they likely were outside of his eyepiece field. This galactic trio lies midway between the stars θ and ι Leonis, and is some 30,000,000 light-years from us (Fig. 7.11).

Orion

H V-32 (NHC 1788): 05 07 – 03 21, diffuse nebula, 11?, 8′ × 5′. *"Bright, considerably large, round, brighter in the middle a triple star of 15th magnitude, a star of 10th magnitude, 1.5' 318 degrees, involved in the nebulosity."* This unusual-looking

Fig. 7.11. H V-8 (NGC 3628) is a very large galaxy lying in the same eyepiece field as M65 and M66. This trio presents a nice sight in even the smallest of telescopes and are referred to collectively as the Leo Triplet. Courtesy of Mike Inglis.

wisp of nebulosity lies just 2°N and slightly W of β Eridani. A 6-inch glass shows it, with a 7.5-magnitude star on its SW edge. There is a fascinating drawing of this object made with a 5-inch telescope that appears in the original *Observe The Herschel Objects* manual, portraying it as a snail-like nebulosity with two unequal stars embedded in it like eyes or nuclei. The brighter one in the middle should be the faint triple Herschel mentioned (unresolved in that drawing) and the dimmer one the 10th-magnitude star (whose position angle and distance are apparently referenced against the central trio). However, the blended image of three 15th-magnitude stars would not outshine that of the second star. Observers using really large apertures may want to verify that these, in fact, comprise the "star" seen at the nebula's center. But there is some uncertainty in Herschel's wording about the triple's brightness and position given in the original *NGC* account above, as well as the reference point for the 10th-magnitude star. As a result, the description of H V-32 that appears in the *NGC 2000.0* has been edited to read: *"Bright, considerably large, round, brighter in the middle triple star; a star of 10th magnitude involved in the nebulosity."*

H V-30 (NGC 1977): 05 36 – 04 52, diffuse nebula, —, 40′ × 20′. *"Remarkable, very much so, c-1 [= 45] 42 Orionis and nebula."* This large dim, elongated nebulosity involving the stars 42 and 45 Orionis lies 30°N of the main mass of the Orion Nebula (M42/M43). It is located between it and the big, sparse open cluster NGC 1981 just N of it (which Herschel apparently did not recognize – see Chapter 12). Overshadowed by its famous neighboring nebulosity, this object is typically neglected (and often unrecognized!) by observers despite the fact that it lies within the same wide-eyepiece field of view with it. From its appearance on photographs, it has become known in recent years as the Running Man Nebula. There are also several fainter stars involved in the nebulosity in addition to 42 and 45 Orionis. A wide field of view, a 4- to 8-inch aperture at low power, and a dark night are all needed to see this object to advantage (Fig. 7.12).

H V-28 (NGC 2024): 05 41 – 02 27, diffuse nebula, —, 30′, Flame/Burning Bush Nebula. *"Remarkable, irregular, bright, very very large, black space included."* Just 15′ E and slightly N of the bright star Alnitak (or ζ Orionis) in Orion's belt is a large nebulosity that surprises you once you realize that it is there! Since both objects lie in the same low-power eyepiece field, you will have to place the star outside of the view if it is not to overwhelm the nebulosity. Medium power on a 4- to 6-inch glass shows its tree-like shape, while larger apertures reveal the dark trunk of the tree or bush as well (the "black space" Herschel mentioned) and lots of complex internal structure. On photographs, this object does indeed look like a flame or burning bush (Fig. 7.13).

Sagittarius

H V-10/11/12 (NGC 6514): 18 02 – 23 02, diffuse nebula, 6.3, 28′, In Messier 20/Trifid Nebula. *"Remarkable, a magnificent or otherwise interesting object, very bright, very large, trifid, double star involved."* As was the case for the supposed planetary nebula H IV-41 (see Chapter 6), here are three more Herschel designations for this well-known Messier object that leave the observer searching for their identity within the nebulosity. The description given above clearly relates to the Trifid itself, and recourse to Sir William's original paper in the *Philosophical Transactions* of the

Fig. 7.12. H V-30 (NGC 1977), sometimes fancifully called the Running Man Nebula from its photographic appearance, is the small horizontal tuft of nebulosity just N of the magnificent Orion Nebula itself. Although Herschel saw the nebulosity, he apparently overlooked the sparse open cluster NGC 1981 just above it. Photographed using a 4-inch apochromatic refractor. Courtesy of Steve Peters.

Fig. 7.13. H V-28 (NGC 2024) is known as the Flame or Burning Bush Nebula. Although a snap to find since it lies in the same wide eyepiece field as ζ Orionis, glimpsing it requires placing the star out of the view so its glare does not overwhelm the dim nebulosity. Photographed using a 4-inch apochromatic refractor. Courtesy of Steve Peters.

Fig. 7.14. H V-10/11/12 refers to three patches of nebulosity cataloged by Herschel within the Trifid Nebula (M20). But actually identifying them at the telescope is definitely a real challenge! Interestingly, all three objects carry the same NGC number (6514) as the Trifid itself. Courtesy of Mike Inglis.

Royal Society would be necessary to clarify what and where his objects are within it. The double star he refers to is known as HN 40 and is actually a triple, consisting of 7th-, 8th-, and 10th-magnitude components just 11″ and 5″ apart. (Apparently the dim star at 5″ was not resolved in his reflectors.) The Trifid is positioned 1.5°N of the much brighter Lagoon Nebula (M8), which is visible to the unaided eye and in binoculars on a dark night above the spout of this constellation's Teapot asterism. Both clouds lie about 5,000 light-years from us (Fig. 7.14).

Sculptor

H V-1 (NGC 253): 00 48 – 25 17, galaxy, 7.1, 25′ × 7′, = Caldwell 65/Sculptor Galaxy. *"Remarkable, very much so, very very bright, very much extended 54 degrees, gradually brighter in the middle."* Generally considered to be the most easily observed spiral after the Andromeda Galaxy (M31), this huge, bright elongated mass lies at the Sculptor-Cetus border, some 7.5°S of the star β Ceti. Such a distance would normally make for a long and uncertain sweep, but this grand galaxy is so big and bold that it is readily visible in finders and binoculars. It was not actually discovered by Sir William but rather by his sister Caroline. And what a discovery! Indeed, an entire chapter could easily be devoted to this marvelous object, which is exciting to observe through any and all size instruments. Despite its rather southerly Declination, even a 2- or 3-inch glass shows it nicely on dark transparent nights. Its oval nucleus and cigar-shaped disk shine brightly in 6- and 8-inch telescopes, which reveal mottling at high powers from its tightly wound spiral arms and dust lanes. Nearly 30′ in length and inclined 12° from being exactly edge-on to our line of sight, this star-city has been aptly described as looking like a tipped dinner plate. This colossus lies 7,500,000 light-years from us. Just 2°SE is the large dim globular cluster H V-16/NGC 288 (see Chapter 6).

Triangulum

H V-17 (NGC 598): 01 34 + 30 39, galaxy, 5.7, 62′ × 39′, = Messier 33/Triangulum/ Pinwheel Galaxy. *"Remarkable, extremely bright, extremely large, round, very gradually brighter in the middle to a nucleus."* Here is another of those perplexing situations in which Herschel assigned a designation to an already existing Messier object – in this case one of the most well-known galaxies in the sky. This huge but somewhat elusive spiral has been amply described elsewhere in countless observing guides. While it has been glimpsed with the unaided-eye and can be seen in binoculars on a dark night, telescopically it is easy to pass right over it in sweeping if you do not know what to expect – a big dim glow about the same apparent size as the full Moon! Once sighted, its magnificence becomes evident – especially in medium-aperture, wide-field telescopes. You will find it almost exactly the same distance SE of the star β Andromedae as the Andromeda Galaxy (M31) is NW of it. For a closer fix, sweep for it 4°W and slightly N of the star α Trianguli. Sir William's description definitely refers to M33 itself rather than to some feature within it (as actually is the case for H III-150 – see Chapter 11). At a distance of 3,000,000 light-years, this starry pinwheel lies slightly further away than M31 itself (Fig. 7.15).

Fig. 7.15. Herschel's description of H V-17 (NGC 598) appears to actually refer to the huge Triangulum or Pinwheel Galaxy itself – which had already been cataloged by Messier and carries his designation M33. It is something of a puzzle why Sir William included it in his catalog. Courtesy of Mike Inglis.

Ursa Major

H V-46 (NGC 3556): 11 12 + 55 40, galaxy, 10.1, 8′ × 2′, = Messier 108. "*Considerably bright, very large, very much extended 79 degrees, pretty bright middle, resolvable (mottled, not resolved).*" This is another of those objects attributed to Messier long after it had been independently discovered by Herschel. It is an attractive edge-on spiral located in a nice field of stars just over 1°SE of β Ursae Majoris in the bowl of the Big Dipper. The famed Owl Nebula (M97) lies in the same wide eyepiece field 48′ to the SE, adding greatly to the overall scene. A 6-inch glass shows both objects nicely. As aperture increases, each object becomes ever-more fascinating – but they can no longer be seen together, as telescopic fields of view correspondingly decrease. This celestial odd couple lies at vastly different distances from each other, the Owl being perched within our galaxy thousands of light-years from us but H V-46 lying far beyond it in the depths of intergalactic space millions of light-years away (Fig. 7.16).

H V-45 (NGC 3953): 11 54 + 52 20, Galaxy, 10.0, 7′ × 4′. "*Considerably bright, large, extended 0 degrees +/–, very suddenly brighter in the middle to a large resolvable (mottled, not resolved) nucleus.*" This spiral is extremely easy to locate, lying in the same eyepiece field as γ Ursae Majoris in the bowl of the Big Dipper, just about 30′ to the SE of the star itself. It is fairly bright and easy to see in a 5- to 6-inch glass, appearing round with an obvious nucleus. Many happy hours can be spent viewing the galaxies discovered by Messier and Herschel within the confines of this large constellation – and contemplating their significance in the grand cosmic scheme of things.

Fig. 7.16. H V-46 (M108) is an edge-on spiral galaxy discovered by Herschel and much later attributed to Messier, who saw it but did not include it in his catalog for some reason. It lies in the same wide eyepiece field as the famed Owl Nebula (M97). Courtesy of Mike Inglis.

Chapter 8

Showpieces of Class VI

Very Compressed and Rich Clusters of Stars

Listed below in alphabetical order by constellation are 20 of the most interesting objects in Herschel's Class VI. Following the Herschel designation itself is the corresponding *NGC* number in parentheses, its Right Ascension and Declination (for Epoch 2000.0), the object's actual type (which may differ from the Class Herschel assigned it to), its visual magnitude, angular size in minutes (′) or seconds (″) of arc, and Messier or Caldwell number plus popular name if any. Next is a translation of Sir William's shorthand description (in italicized quotes) taken from the *NGC* itself, followed by comments from the author. These include directions for finding each object by sweeping for it, just as Herschel himself originally did.

Bootes

H VI-9 (NGC 5466): 14 06 + 28 32, globular cluster, 9.1, 11′. *"Cluster, large, very rich, very much compressed, stars from the 11th magnitude downwards."* This rather dim fuzz-ball lies 2°NE of the star 9 Bootis. Interestingly, it can also be found by sweeping 4° due E of another much brighter cluster – the well-known spring globular M3! It looks like a round nebula in small telescopes, in striking contrast to the stellar appearance of its famous neighbor. Although Herschel did not actually mention resolving this object, he did see stars within it and considered it to be a considerably rich cluster. Apertures of at least 10-inches or more, high magnification, and good seeing are all needed to view it as the stellar beehive that it actually is (Fig. 8.1).

Cassiopeia

H VI-31 (NGC 663): 01 46 + 61 15, open cluster, 7.1, 16′, = Caldwell 10. *"Cluster, bright, large, extremely rich, stars pretty large."* This sparkling clan lies at the center of a trio of Herschel star clusters positioned within the same wide eyepiece field. The other two objects are H VII-46 (NGC 654) and H VIII-65 (NGC 659), both of which are much smaller and fainter than H VI-31. You will find them at the E

Fig. 8.1. HVI-9 (NGC 5466) is a relatively dim globular cluster largely unnoticed by observers due both to its intrinsic obscurity and to its being completely overshadowed by the bright showpiece globular M3 just a few degrees away. Courtesy of Mike Inglis.

corner of a triangle formed with δ and ε Cassiopeiae, two of the stars in the familiar "W" (or "M") shape of this constellation. There are some 80 suns in this clan visible in 6- and 8-in. apertures. Only a handful of the brighter ones stand out individually against an unresolved background of fainter stars in small telescopes. Included within the cluster are three rather faint double stars – Struve 151, Struve 152 and Struve 153 – all 9th- and 10th-magnitude pairs having separations between 7″ and 8″.

H VI-30 (NGC 7789): 23 57 + 56 44, open cluster, 6.7, 16′, Caroline's Cluster. "*Cluster, very large, very rich, very much compressed, stars from the 11th to 18th magnitude.*" The author named this object in honor of its discovery by Caroline Herschel. Easily found positioned about midway between the stars ρ and σ Cassiopeiae, this remarkably rich and uniform-looking cluster contains at least 300 suns and its total membership is thought to be in excess of 900. (Note that Herschel called it "very much compressed.") It can be seen in a 2-inch glass on a dark night and is a splendid sight in a 6- or 8-inch, where stars against stardust (its unresolved fainter members) are seen. Given a large enough field of view, its appearance in 12- to 14-inch scopes is truly magnificent! (To many observers including the author, this splendid swarm seems to be nearly twice the apparent size assigned to it.) Overlooked by Messier, it was also surprisingly bypassed for the Caldwell listing. This lovely star-cloud is intermediate in concentration between the rich open clusters and the loose globulars, and may actually be another of those objects like Smyth's Wild Duck Cluster (M11) in Scutum that appear to be "semi-globular" in nature. This softly glowing stellar commune lies 6,000 light-years from us (Fig. 8.2).

Cepheus

H VI-42 (NGC 6939): 20 31 + 60 38, open cluster, 7.8, 8′. "*Cluster, pretty large, extremely rich, pretty compressed in the middle, stars from the 11th to 16th magnitude.*" This rather dim, tight collection of stars contains between 80 and 100 members

Fig. 8.2. H VI-30 (NGC 7789) is a beautiful rich swarm of over 300 suns having a very uniform appearance in the eyepiece. This stellar jewelbox is among those objects in the Herschel catalog discovered by Caroline Herschel rather than by Sir William himself. Courtesy of Mike Inglis.

concentrated within a very small area of sky. Located about 2°SW of η Cephei at the Cygnus border, it is a misty-looking group as seen in a 4- to 6-inch glass and really needs aperture to transform it into the rich stellar clan it is. Adding to the interest of the scene here is the galaxy H IV-76/NGC 6946 (see Chapter 6) lying in the same wide eyepiece field 38′ to the SE. Although this unique combo appears close together in the sky, the two objects are actually vastly far apart in space – the galaxy being some 5,000 times the distance of the cluster (10,000,000 light-years compared to 2,000 light-years). Talk about "extreme depth of field" in the eyepiece! Note that Sir William did not mention about seeing these two objects together (which is really the big attraction here), the galaxy apparently lying outside the limited fields of view of his large telescopes (Fig. 8.3).

Coma Berenices

H VI-7 (NGC 5053): 13 16 + 17 42, globular cluster, 9.8, 10′. *"Cluster, very faint, pretty large, irregularly round, very gradually brighter in the middle, stars 15th magnitude."* This ghostly starball is neatly positioned just a degree SE of the prominent globular cluster M53, which itself is 1.5°NE of the star α Comae Berenices. At least an 8-inch telescope is needed to make out this object most nights and larger apertures are an absolute must in order to see any real detail. In 12- to 14-inch instruments this object's unusual appearance becomes evident, looking like either a rich open cluster or a very loose globular. And so dim is it that one observer has described H VI-7 in the eyepiece as being the departed soul of its more brilliant neighbor (Fig. 8.4)!

Fig. 8.3. H VI-42 (NGC 6939) is a fairly dim and tight collection of stars, which by itself is not overly impressive. But lying within the same wide eyepiece field is the face-on spiral galaxy H IV-76 (NGC 6946). Together, they offer a fascinating contrast not only in visual appearance but also in distance – the galaxy lying thousands of times further from us than the cluster. Courtesy of Mike Inglis.

Fig. 8.4. HVI-7 (NGC 5053) is a dim globular cluster lying just 1°SE of the much brighter starball M53. Large apertures and dark transparent nights are needed to show it to advantage. A wide field of view includes both objects, which present a fascinating contrast to each other. Courtesy of Mike Inglis.

Gemini

H VI-17 (NGC 2158): 06 08 + 24 06, open cluster, 8.6, 5′. *"Cluster, pretty small, much compressed, very rich, irregular triangle, stars extremely small."* This fascinating little cloud of stars lies on the SW edge of the large radiant open cluster M35 (which the author has dubbed Lassell's Delight from that classic English observer's excited account of it!) within the same eyepiece field. You will find this dual stellar commune about 2°NW of the star η Geminorum in a rich Milky Way field,

M35 itself being readily seen in a finder. Visible in 3- and 4-inch glasses as a gentle glow located just beyond the big cluster's outlying stars, it sparkles as seen in 5- and 6-inch scopes using averted vision. Larger apertures resolve many of its more than 150 stars, and show it actually looking more like a globular cluster than an open one (something that is especially evident on photographs). Indeed, some sources consider this to be another of that class of "semi-globulars" – a possible transition between the two types. This clan seems much fainter in the eyepiece than is suggested by the brightness given above, some observers even rating it as dim as 11th-magnitude (which it definitely is not). While these star-swarms appear close together in the eyepiece, they are actually very far apart in space. M35 lies 2,700 light-years from us while H VI-17 is on the order of 16,000 light-years away – very remote for an open cluster but a distance typical of that of many globular clusters (Fig. 8.5).

H VI-21 (NGC 2266): 06 43 + 26 58, open cluster, 9.8, 7′. *"Cluster, pretty small, extremely compressed, rich, stars from the 11th to 15th magnitude."* This small, dim group is being offered as an example of those Herschel objects which are so neglected that they are not shown on most star atlases – and yet they lie within reach of 4- to 8-inch telescopes. In the case of H VI-21, the three widely used star maps *Norton's 2000.0*, *The Cambridge Star Atlas* and *Sky Atlas 2000.0* all fail to plot it. However, Sky Publishing's new, superbly executed *Pocket Sky Atlas* shows it. Sweep for it about 2° due N of the close unequal double star ε Geminorum. Here you will find several dozen stars tightly packed into a very tiny area, which some sources give as just 5′ in size rather than the 7′ listed here. One fairly bright star stands out from among the rest and the group has a triangular shape to it. Large apertures offer a fascinating view of this overlooked stellar clan.

H VI-1 (NGC 2420): 07 39 + 21 34, open cluster, 8.3, 10′. *"Cluster, considerably large, rich, compressed, stars from the 11th to 18th magnitude."* Here is a neat, misty-looking starry commune to check out when viewing the nearby Eskimo/Clown Face Nebula (H IV-45/NGC 2392 – see Chapter 6), which lies 2°SW of the cluster itself. Sweep for it 3°SW of the star κ Geminorum, which lies beneath

Fig. 8.5. H VI-17 (NGC 2158) lies on the SW outskirts of the big, bright splashy open cluster M35. Although classified as an open cluster as well, it looks more like a globular both visually and on photographs. Lying well beyond the stars of M35 itself some five times as remote, its distance is also more typical of a globular than that of an open cluster. Courtesy of Mike Inglis.

Pollux. A 6-inch glass shows this group nicely, with more than a dozen suns shining amid a background of much fainter members – all encompassed by several brighter field stars.

Hydra

H VI-22 (NGC 2548): 08 14 – 05 48, open cluster, 5.8, 30′, = Messier 48. "*Cluster, very large, pretty rich, pretty much compressed, stars from the 9th to 13th magnitude.*" This big, bright splash of some 50 stars is a nice sight in a 2- or 3-inch glass and is quite striking in larger scopes, providing they have a wide field of view. Covering an area the same apparent size as the full Moon, its members are arranged roughly in the shape of a triangle. This is one of the infamous "missing" Messier objects – which was lost and then found! It was independently discovered by Caroline Herschel. Easily picked up by sweeping 3.5°SW of the little naked-eye clump of stars formed by C, 1 and 2 Monocerotis, it lies 1,900 light-years from us.

Libra

H VI-19 = H VI-8? (NGC 5897): 15 17 – 21 01, globular cluster, 8.6, 13′. "*Globular cluster of stars, pretty faint, large, very irregularly round, very gradually brighter in the middle, well resolved, clearly consisting of stars.*" Despite its rather pale-looking appearance, this starball is generally considered to be the deep-sky showpiece of Libra (which is surprisingly empty of nonstellar deep-sky wonders, the others being a handful of faint galaxies!). Sweep for it about 1.5°SE of the unequal, tight multiple star ι Librae. It is a large but extremely low surface brightness globular, whose members are uniformly distributed with no obvious central concentration like that typical of most members of its class. Its appearance caused Sir William (stated elsewhere) to regard this object as a transition between a cluster and a nebula – which at that time were all considered to be simply unresolved masses of stars. And here is another of those objects that seem to be intermediate in nature between the rich open clusters and weak globulars. This unusual clan needs at least an 8- to 10-inch aperture and a dark, steady night to resolve. Seen in smaller scopes, it looks faint, nebulous and remote. Some sources give its brightness as between 10th- and 11th-magnitude – and while it certainly does appear to be fainter in the eyepiece than the magnitude given above would suggest, this is mainly a result of its large apparent angular size. According to the *NGC*, H VI-19 is thought to be the same object as Herschel's H VI-8, both appearing under the same number with the "?" notation shown above.

Monoceros

H VI-27 (NGC 2301): 06 52 + 00 28, open cluster, 6.0, 12′. "*Cluster, rich, large, irregular figure, stars large and small.*" This stellar jewelbox lies 5°W and 30′ N of the star-pair δ and 24 Monocerotis, in a very rich region of the Winter Milky Way.

It hosts some 60 suns in an area less than half the apparent size of the full Moon. Herschel remarked about its irregular shape, which some observers have likened to a bird in flight. It is a nice sight even in a 3- or 4-inch glass, and its appeal grows with aperture.

H VI-37 (NGC 2506): 08 00 – 10 46, open cluster, 7.6, 12', = Caldwell 54. "*Cluster, pretty large, very rich, compressed, stars from the 11th to 20th magnitude.*" This lovely clan is both compact and rich, containing some 50 uniform-looking stars as seen in 6-inch or smaller scopes and nearly twice that number in large backyard instruments. As so often happens in the case of rich open clusters, in small glasses this object appears to contain nebulosity (actually being its unresolved fainter members) against which its brighter stars are sprinkled. Sweep for it 5°SE of α Monocerotis, at the Monoceros–Puppis border.

Ophiuchus

H VI-40 (NGC 6171): 16 32 – 13 03, globular cluster, 8.1, 10', = Messier 107 "*Globular cluster of stars, large, very rich, very much compressed, round, well resolved, clearly consisting of stars.*" Although Herschel described this rich stellar beehive as being large, it actually appears rather small in typical amateur-class telescopes. Indeed, apparent angular sizes of somewhere between 2' and 4' are often reported here, despite the catalog value listed above. In any case, at least a 12-inch aperture is needed to clearly resolve its host of faint suns. It has a definite grainy appearance (the precursor of actual resolution) in a 6-inch at high power while in smaller scopes it looks like a round "nebula" with a bright center. It is easily swept up 2.5°S and slightly W of the star ζ Ophiuchi. H VI-40 is another of those objects attributed to Messier long after Sir William had independently discovered it (Fig. 8.6).

Fig. 8.6. H VI-40 (M107) is a very rich but remote globular cluster discovered by William Herschel and later attributed to Messier, who apparently saw it but failed to include it in his original catalog. Courtesy of Mike Inglis.

Orion

H VI-5 (NGC 2194): 06 14 + 12 48, open cluster, 8.5, 10′. *"Cluster, large, rich, gradually very much compressed in the middle."* Here is a nice, rich starry clan located about 1.5°S and slightly E of ξ Orionis, sitting right up against the stars 73 and 74 Orionis. Containing some 100 members in an area actually somewhat smaller in size than the 10′ given above, it looks quite concentrated in the eyepiece at medium magnifications. Indeed, an aperture of at least 6- or 8-inches is required to see it as a glittering stellar commune. On short exposure photographs, it has the appearance of the letter "H" and something of this can also be made out visually as well. In any case, a definite "bar" of stars can be seen crossing its center. This tightly knit clan is perhaps the best of the open clusters to be found in this rich constellation (which the French astronomer Flammarion aptly called the "California of the Sky").

Perseus

H VI-33 (NGC 869): 02 19 + 57 09, open cluster, 3.5, 30′, = Caldwell 14/Double Cluster. *"Remarkable, cluster, very very large, very rich, stars from the 7th to 14th magnitude."*/**H VI-34 (NGC 884): 02 22 + 57 07, open cluster, 3.6, 30′, = Caldwell 14/Double Cluster.** *"Remarkable, cluster, very large, very rich, ruby star in the middle."* Just how or why Charles Messier overlooked this dazzling pair of stellar jewelboxes is definitely something of an enigma! The argument that they are obviously not comets and so he did not bother to catalog them does not solve the mystery, since he *did* include both the Beehive (M44) and Pleiades (M45) naked-eye clusters. The Double Cluster is one of those celestial showpieces to which an entire chapter could easily be devoted, but just an overview of its wonders is given here. Let us begin by finding this dual clan. They are easily spotted at the SE corner of a triangle formed by the stars ε and δ Cassiopeiae. They also lie midway between the latter star and γ Persei. The spectacle greeting the observer here is wondrous and stunning, even in a 3-inch glass! Both clusters fit nicely within the same wide eyepiece field (at least 1.5° is preferred, and greater if possible – the more sky encompassing them, the more dramatically they stand out from the rich Milky Way background here), their cores being about 30′ apart. Together they rank as the finest object of their class for small telescopes after the classic Pleiades themselves for N. Hemisphere observers. In larger instruments, typically only one at a time can be seen – unless they happen to be short-focus, rich-field telescopes (RFTs – which most big Dobsonian reflectors are), in which case a totally awesome sight overwhelms the unprepared observer. This is especially true should a binocular eyepiece be used, causing these radiant communes to appear hanging suspended against the Milky Way background! H VI-33 (the W member) is the richer of the two, containing an amazing "star tunnel" – a term that the author coined from what is seen gazing into its glittering core. Can you see it? Membership is variously estimated at between 200 and 350 for H VI-33 and from 150 to 300 for H IV-34. Most are bluish-white in hue, but both clusters contain a number of red and orange (mainly variable) stars, including the ruby one in the center of H VI-34 mentioned by Herschel. There is also a very red star positioned between the two objects where their outer regions overlap. Incidentally, in case you are wondering if this duo is really physically (gravitationally)

Fig. 8.7. H VI-33 (NGC 869) & H VI-34 (NGC 884) together form the magnificent Double Cluster. Visible to the unaided eye and a fascinating sight even in binoculars, it remains a real mystery why Messier did not include these glittering stellar jewelboxes in his catalog – but fortunately, Sir William did! Photographed using a 300-mm *f*/2.8 camera. Courtesy of Steve Peters.

related, they are! The distance of both from us is nearly identical – 7,300 light-years, with an uncertainty of just +/–170 light-years. It is also believed that they are slowly orbiting around their common center of gravity in a period measured in millions of years, like some colossal "binary star." And finally, do not overlook viewing the Double Cluster through a pair of binoculars – in even the smallest of such humble glasses, it is a lovely sight (Fig. 8.7)!

H VI-25 (NGC 1245): 03 15 + 47 15, open cluster, 8.4, 10′. *"Cluster, pretty large, rich, compressed, irregularly round, stars from the 12th to 15th magnitude."* This cloud of more than a hundred minute stars is positioned on the W edge of the famed Alpha Persei Association – a big bright naked-eye splash of stellar gems surrounding Mirfak (α Persei) and cascading downward in the sky from it. The cluster itself contains a number of brighter stars that stand out against the uniform glow of its fainter members and they are what is typically seen (at least on first glance) in 2- to 4-inch glasses. Apertures in the 6–8 inch range pick up the dimmer stars and nicely fill in the cluster. Surrounding it all are several field stars that are arranged in the shape of a foreshortened pentagon. The Alpha Persei Association itself is only about 600 light-years from us, but H VI-25 appears to lie far beyond it at a much greater distance.

Sagittarius

H VI-23 (NGC 6645): 18 33 – 16 54, open cluster, 8.5, 10′. *"Cluster, pretty large, very rich, pretty compressed, stars from the 11th to 15th magnitude."* This little stellar commune can be found by sweeping 2.5°SE of γ Scuti (which is thrilling to do in this incredibly rich region of the Milky Way!). Some 40–60 mostly dim stars are seen here, depending on the size of telescope used – this object definitely needing larger apertures to be appreciated. Short exposure photographs show two apparent rings of stars surrounding central vacancies or holes, and these can be picked out visually upon careful inspection.

Scorpius

H VI-10 (NGC 6144): 16 27 – 26 02, globular cluster, 9.1, 9'. *"Cluster, considerably large, much compressed, gradually brighter in the middle, well resolved, clearly consisting of stars."* This tiny dim stellar beehive is one of the most fascinating of all the Herschel clusters to be observed due to its unique setting. Positioned just 30' NW of fiery Antares (α Scorpii) and lying in the same wide eyepiece field with it, this object is a snap to locate but a challenge to see in most telescopes. The problem here is glare from that radiant supersun flooding the field of view. Placing Antares just outside the edge of the eyepiece helps, but its overpowering presence is still noticeable. Some sources list this little (Herschel called it "considerably large"?) ball as just 3' in size and a magnitude fainter than the catalog values given above – both of which better match what is actually seen in the eyepiece. It can be glimpsed in 4- to 6-inch telescopes on a steady, transparent night but at least a 12-inch is needed to even begin resolving its tightly packed suns. While here, do not miss viewing the magnificent globular cluster M4, located 1° to the SW of H VI-10 and just over 1°W of Antares. What a contrast this glittering swarm of stars (which can be resolved to its core even in small telescopes) provides to its faint neighboring cluster! Antares and both starballs can be seen all together in the same telescopic view, given a field at least 1.5° in extent. (Also, Antares itself is a beautiful reddish-orange and emerald-green tight double for steady nights in 5-inch and larger telescopes.)

Sculptor

H VI-20 (NGC 288): 00 53 – 26 35, globular cluster, 8.1, 14'. *"Globular cluster of stars, bright, large, little extended, stars from the 12th to 16th magnitude."* Here is a nice large but inherently dim globular that is often overlooked by observers, due mainly to its rather low Declination. Yet it is located just 1.5°SE of the big, bright spiral H V-1/NGC 253 (see Chapter 7); better-known as the Sculptor Galaxy, this showpiece *is* widely observed! You can also find H VI-20 by sweeping 3°N and slightly E of the star α Sculptoris. Visible in binoculars, it appears as a round unresolved haze in a 3- or 4-inch glass. Despite its relative faintness, its stars are loosely packed and easily resolved at medium magnification on 8-inch and larger telescopes. In 12- to 14-inch apertures, it glitters to its core on dark, steady nights! Even though some sources consider this object to be as bright as 7th-magnitude, its light is spread out over a fairly large area, accounting for its pale aspect (along with its low attitude in the sky). Note that for some reason Herschel did not specifically mention resolving this cluster, even though he did clearly see stars within it as stated.

Chapter 9

Showpieces of Class VII

Compressed Clusters of Small and Large Stars

Listed below in alphabetical order by constellation are 18 of the most interesting objects in Herschel's Class VII. Following the Herschel designation itself is the corresponding *NGC* number in parentheses, its Right Ascension and Declination (for Epoch 2000.0), the object's actual type (which may differ from the Class Herschel assigned it to), its visual magnitude, angular size in minutes (') or seconds (") of arc, and Messier or Caldwell number plus popular name if any. Next is a translation of Sir William's shorthand description (in italicized quotes) taken from the *NGC* itself, followed by comments from the author. These include directions for finding each object by sweeping for it, just as Herschel himself originally did.

Andromeda

H VII-32 (NGC 752): 01 58 + 37 50, open cluster, 5.7, 50′, = Caldwell 28. "*Cluster, very very large, rich, stars large and scattered.*" This sprawling stellar jewel box contains more than 60 suns scattered over an area much larger than the full Moon. It is visible in binoculars and is a lovely sight in 3- to 6-inch telescopes at very low powers. Larger instruments typically cannot fit the entire cluster within the field of view – unless they happen to be wide-angle, rich-field telescopes (RFTs). Given the relatively narrow fields of Herschel's large reflectors, it is doubtful that he saw this lovely commune in anything like its full glory. "Big, sparse, scattered and bright" best describes H VII-32. It is easily found by sweeping 4°S and slightly W of the beautifully-tinted double star Almach (γ Andromedae). Adding to the spectacle here is the wide matched orange double star 56 Andromedae (magnitudes 5.7, 5.9, and separation 190″) parked on the SW edge of the cluster. The two objects make a great combo but are physically unrelated, lying at distances of 360 light-years and 1,200 light-years, respectively (Fig. 9.1).

Fig. 9.1. H VII-32 (NGC 752) is a big scattered open cluster best seen in rich-field telescopes (RFTs). From its size, this extensive stellar clan seems to more properly belong to Herschel's Class VIII rather than VII. And it is definitely anything but compressed! The wide double star mentioned in the text lies outside the field of this image. Courtesy of Mike Inglis.

Aquila

H VII-19 (NGC 6755): 19 08 + 04 14, open cluster, 7.5, 15′. *"Cluster, very large, very rich, pretty compressed, stars from the 12th to 14th magnitude."* For being situated smack in a rich part of the Milky Way, Aquila has very few star clusters – either globular or open. One of the best of the latter is NGC 6709, which Herschel missed (see Chapter 12). But another one he did discover is H VII-19. It is located about 4.5°NW of δ Aquilae. A line drawn from the well-known variable star η Aquilae extended through δ points right at it. Here we find several dozen stars loosely scattered over an area half the size of the full Moon. A 4-inch glass shows it but it looks much nicer in an 8-inch, which brings out its fainter members.

Auriga

H VII-33 (NGC 1857): 05 20 + 39 21, open cluster, 7.0, 6′. *"Cluster, pretty rich, pretty compressed, stars from the 7th magnitude downwards."* In contrast with Aquila (see above), Auriga is loaded with open star clusters. Most of them, however, are overshadowed by the well-known trio of Messier objects M36, M37, and M38. But many others are worth having a look at. While certainly not a spectacular sight, little H VII-33 hosts some four dozen stars within its confines. And the attraction here is that a number of them are as bright as 7th- to 8th-magnitude, which is typically not the case for most non-Messier open clusters. This tiny stellar commune definitely benefits from aperture, at least a 6-inch being needed to appreciate it.

Camelopardalis

H VII-47 (NGC 1502): 04 08 + 62 20, open cluster, 5.7, 8′, Golden Harp Cluster. *"Cluster, pretty rich, considerably compressed, irregular figure."* Situated about 2° due N of the planetary known as the Oyster Nebula (H IV-53/NGC 1501 – see Chapter 6), this little clan of several dozen suns has an obvious trapezoidal shape giving rise to its unusual name. (Some observers see it as a cross-bow or an irregular triangle rather than a harp.) Included among its membership are the dim multiple stars Struve 484 (all 10th-magnitude at distances of 5″, 23″ and 49″) and Struve 485 (magnitudes 7, 7, 10 and 10 at distances of 18″, 70″ and 139″). However, distinguishing their faint components from the cluster's other stars can be a real problem here. This object is easy to see but hard to locate due to its isolation from any bright guide stars around it. Sweeping nearly 1 h of Right Ascension (or 15°) due W of the star β Camelopardalis brings you to the neighboring planetary, and then just a short hop N lands you on the cluster itself. It is best viewed in 8-inch and larger apertures at medium magnifications (Fig. 9.2).

Fig. 9.2. H VII-47 (NGC 1502) is popularly known as the Golden Harp Cluster from the unique irregular arrangement of its stars. As is the case for most open clusters, such patterns are typically more obvious in the eyepiece than they are on photographs. Courtesy of Mike Inglis.

Canis Major

H VII-12 (NGC 2360): 07 18 – 15 37, open cluster, 7.2, 13′, = Caldwell 58. "*Cluster, very large, rich, pretty compressed, stars from the 9th to 12th magnitude.*" Here is a lovely cluster of some 50–60 suns sitting in a rich Milky Way region just 3.5° due E of γ Canis Majoris. It is an attractive sight in a 4- or 5-inch glass, its sparkling gems appearing fairly uniform-looking with a few brighter ones evident. There is also a 6th-magnitude star sitting on the W edge of the cluster. Some observers have remarked that this object seems to "melt" into the rich stellar background of our Galaxy. It is another lovely discovery of Caroline Herschel rather than being one found by Sir William himself.

H VII-17 (NGC 2362): 07 19 – 24 57, open cluster, 4.1, 8′, = Caldwell 64/Tau CMA Cluster. "*Cluster, pretty large, rich (30 Canis Majoris).*" Here is a gorgeous little cluster that must surely have surprised and thrilled Sir William when he chanced upon it – just as it does observers today once they realize what they are seeing! This tight clan is easily located since it surrounds the 4th-magnitude blue-giant sun τ Canis Majoris (which is actually a cluster member and not just a foreground star!). At low magnifications on small scopes, it is easy to overlook this group, since it is lost in the glare of its central sun. But from medium to high powers on 6-inch and larger instruments, the cluster members suddenly appear like a "starburst of fireworks frozen in space" (as one observer excitedly described it), encompassing glittering τ itself. The overall effect here is like seeing some exquisite piece of cosmic jewelry! And yet this glittering diamond jewel box is seemingly little-known and seldom observed. H VII-17 is one of the very best Herschel clusters and, indeed, one of the most exciting of all his discoveries. If you have not yet seen it, you simply must do so the next clear night – even if this means getting up before dawn to catch the Winter sky. You will not be disappointed! It sparkles at us from across 5,400 light-years of interstellar space. And if you are a double star lover, while here sweep about 2°N and slightly W to the magnificent Winter Albireo (another name coined by the author from its resemblance to that famous orange and blue pair). This lovely combo was discovered by Sir John Herschel and carries his designation h3945.

Cassiopeia

H VII-42 (NGC 457): 01 19 + 58 20, open cluster, 6.4, 13′, = Caldwell 13/Owl/ET Cluster. "*Cluster, bright, large, pretty rich, stars of 7th, 8th and 10th magnitude.*" This striking group of some 80 suns is arranged in the shape of an owl, with golden φ Cassiopeiae and a nearby 7th-magnitude star marking its big bright eyes! Both of these "eyes" are supergiants, and even though they lie out at the SE edge of the cluster they are believed to be physical members. Other names given to this group in recent years are the ET Cluster and the Dragonfly Cluster. A 3-inch glass shows it nicely and it is a grand sight in 6-inch and larger scopes at low magnification. This delightful clan is easily found, occupying the S corner of a triangle formed with γ and δ in the "W" (or "M") of Cassiopeia. It is considered to be one of the best clusters missed by Messier. The distance given for φ and its attending group is a remote 9,300 light-years.

H VII-48 (NGC 559): 01 30 + 63 18, open cluster, 9.5, 7′, = Caldwell 8. "*Cluster, bright, pretty large, pretty rich.*" This little cluster lies at the NW apex of an equilateral triangle formed with the stars δ and ε Cassiopeiae in the familiar "W" (or "M") of this distinctive constellation. With more than 50 stars visible in medium to large amateur instruments, this rich group is on the small and dim side. And for its size, it is also not very concentrated. (Some sources rate this object as two full magnitudes brighter than the value given here and only about half as big, but what is seen in the eyepiece does not support this.) There is one brighter star evident within the cluster, as well as a number of apparent chains of stars – something that is quite commonly seen with open clusters in general. These are believed to be chance alignments rather than actual physical streams of stars, but this is still not certain. Real or illusory, star-chains can be quite striking in some cases once the eye manages to pick them out.

Cepheus

H VII-44 (NGC 7510): 23 12 + 60 34, open cluster, 7.9, 4′. "*Cluster, pretty rich, pretty compressed, fan-shaped, stars pretty bright.*" Lying near the Cepheus–Cassiopeia border (and sitting precisely on the galactic equator), this unusual-looking group can be found by sweeping just over 1°NE of the star-pair 1 and 2 Cassiopeiae. Here we find several dozen suns arranged in a triangular or arrowhead shape as seen in a 3- to 6-inch glass, and as a kind of thin parallelogram in larger scopes as the cluster's fainter members come into visibility. Note that Herschel described it as fan-shaped. In any and all cases, this certainly is not your typical symmetrically round open star cluster! And following up on our discussion of star-chains in H VII-44 above, there is a couple of dramatic ones to be seen here among this clan's fainter members – given sufficient aperture to pick them out.

Cygnus

H VII-59 (NGC 6866): 20 04 + 44 00, open cluster, 7.6, 7′, The Kite Cluster. "*Cluster, large, very rich, considerably compressed.*" Sweeping through the Cygnus Milky Way at a point 3°SE of the tight double star δ Cygni, in the crossarm of this constellation's Northern Cross asterism, you will encounter another distinctively shaped cluster. Its members are arrayed in a way that reminds many observers of a celestial kite flying against a glistening backdrop of stars! With more than 40 fairly bright stellar pinpoints visible in a 6- or 8-inch telescope, this group appears rich for its size despite the fact that it is officially classified as a loose cluster. Note that Herschel himself described this clan as very rich and compressed.

Monoceros

H VII-2 (NGC 2244): 06 32 + 04 52, open cluster, 4.8, 24′, = Caldwell 50/Rosette Cluster. "*Cluster, beautiful, stars scattered (12 Monocerotis).*" Positioned within the central hole of the huge Rosette Nebula, we find this large, bright, and loosely packed cluster of more than a dozen suns centered on the yellow giant 12 Monocerotis. Sweep for this fascinating dual object just over 2°E and slightly N of the beautiful double star ε (=8) Monocerotis. The faint ring of nebulosity itself is more than twice the apparent size of the full Moon. Herschel missed it (see Chapter 12) and so have many other observers, who were looking right through the central opening without realizing it! Large binoculars and RFTs in the 4- to 6-inch aperture range show both the nebula and the cluster on a dark night. And while the former is somewhat of a challenge to see in typical telescopes with their relatively limited fields of view, the latter is readily seen in the smallest of glasses and has even been glimpsed with the unaided eye. The brighter stars mentioned above are arranged in a kind of rectangular pattern, which is not exactly centered on the hole itself. In 10-inch and larger backyard telescopes, more than 90 fainter stars have been counted visually scattered within and around in the rectangle. Both the cluster and nebula lie 2,600 light-years from us, buried within the Winter Milky Way.

Perseus

H VII-61 (NGC 1528): 04 15 + 51 14, open cluster, 6.4, 24′. *"Cluster, bright, very rich, considerably compressed."* Despite having some 80 members, this clan at first glance seems sparse for its size. And for this reason, it may actually seem more appealing in small glasses at low powers than in larger telescopes. But closer inspection with 8-inches aperture or more reveals its true richness. Sweep for it about 1.5°NE of λ Persei. Be careful here not to confuse it with another nearby Herschel cluster, H VIII-85 (NGC 1545), which lies 2° due E of λ. It is fainter, smaller, and has fewer stars than does H VII-61. (It also has two dim naked-eye beacons parked on its edge in the direction of λ.) As in the case of H VI-25 (NGC 1245 – see Chapter 8) and other Herschel clusters in Perseus, this object is completely overshadowed by the presence of the radiant Double Cluster and also the lovely stellar jewel box M34.

Puppis

H VII-11 (NGC 2539): 08 11 – 12 50, open cluster, 6.5, 22′. *"Cluster, very large, rich, little compressed, stars from the 11th to 13th magnitude."* This stellar commune hosts some 100 fairly dim and uniform-looking suns scattered over a sizeable area of sky. There is one brighter star near its center, but its crown jewel is actually the multiple star 19 Puppis (magnitudes 4.7, 8.9 and 7.8 at separations of 60″ and 71″). Parked on the SE edge of the cluster, it makes finding this group relatively easy while adding greatly to the scene here. H VII-11 also lies 7°S and just slightly W of the big, bright open cluster M48 (H VI-22/NGC 2546 – see Chapter 8). It is best viewed in 6-inch and larger apertures due to the relative faintness of its members.

H VII-64 (NGC 2567): 08 19 – 30 38, open cluster, 7.4, 10′. *"Cluster, pretty large, pretty rich, little compressed, irregularly round, stars from the 11th to 14th magnitude."* This stellar clan needs a clear horizon and transparent night to be seen well due to its low Declination. And as for all celestial wonders – but especially those at low elevations – observing it when on the meridian (and, therefore, at its highest in the sky) is essential for an optimum view. Here a 6-inch glass shows some two dozen mostly faint stars with a few brighter ones mixed in and scattered about, all set in a rich Milky Way field. Larger apertures significantly increase this number, revealing this to be a fairly rich cluster. Finding H VII-64 can be a bit of a challenge, due to both the absence of nearby bright field stars and the presence of several other Herschel clusters in the area. Sweep for it about midway along a line from β to ρ Puppis. This will most likely bring H VI-39 (NGC 2571) into view first, with H VII-64 lying less than 1° to its S. Both clusters will actually fit in the same low-power eyepiece view, given a field of 1.5°.

Sagittarius

H VII-7 (NGC 6520): 18 03 – 27 54, open cluster, 8.1, 6′. *"Cluster, pretty small, rich, little compressed, stars from the 9th to 13th magnitude."* Sweeping 2.5°N and slightly W of the star γ Sagittarii in the spout of this constellation's Teapot asterism bring us to a most fascinating object. (Note that there are two very small faint globular clusters just 3′ apart lying in the same eyepiece field as γ, immediately NW of the star. The more obvious one is H I-49/NGC 6522 and the other is H II-200/NGC 6528.) H VII-7 contains some two dozen stars nicely packed into a relatively small area, giving it a rich appearance and offering a lovely sight in 6-inch and larger telescopes. But the real attraction here is the prominent dark nebula Barnard 86 positioned just off its W edge. This apparent "hole in the sky" measures 4′ × 3′ in extent and is flanked by stars on its opposite sides. It is quite distinct even in a 4-inch glass at medium magnification. The stark absence of stars due to obscuration by this black dust cloud offers a dramatic contrast to the star-rich cluster next to it! This combo (which is believed to actually be physically related) lies over 6,000 light-years from us in a gloriously rich part of the Sagittarius Milky Way, adding greatly to the overall effect.

Taurus

H VII-21 (NGC 1758): 05 04 + 23 49, open cluster, 7.0, 42′. *"Cluster, pretty compressed, stars large and small."* This is one of two "dual clusters" in Taurus (see also next), which in both cases Herschel cataloged the one but not the other! What is taken to be H VII-21 appears as a large but sparse arc of fairly bright stars. However, this is just the E section of the much bigger and richer attached cluster NGC 1746. Sir William must have seen this object but neither described it (although his description above leaves some doubt as to which group he was actually referring to – perhaps both combined?) nor gave it a designation. On virtually all modern star atlases, NGC 1746 is plotted without NGC 1758 – with the exception of the many early editions of the original *Norton's Star Atlas*, which show H VII-21 but not NGC 1746! The *NGC* itself tersely describes the latter object as a "cluster, poor" despite the fact that it contains some 50 stars from 8th-magnitude and downward. Interestingly, *Sky Atlas 2000.0* does show several bright stars in an arc at the edge of the cluster symbol for NGC 1746, which is supposed to be

H VII-21. This enigmatic duo is perhaps best seen in medium aperture telescopes under dark sky conditions. Look for it just slightly S of the midpoint of a line joining the stars 99 and 103 Tauri, within the horns of the celestial bull. It also lies on a line connecting the radiant orange sun Aldebaran (α Tauri) and β Tauri (a star that is actually shared with the adjoining constellation of Auriga), being somewhat closer to the latter.

H VII-4 (NGC 1817): 05 12 + 16 42, open cluster, 7.7, 16′. "*Cluster, large, rich, little compressed, stars from the 11th to the 14th magnitude.*" This is the second of the two "dual clusters" to be found in Taurus (see above). Medium apertures show H VII-4 itself as a fairly large and loose scattering of more than a dozen obvious stars along with many fainter ones. Lying in the same wide eyepiece field and overlapping it is NGC 1807, which was apparently not recognized by Herschel (see Chapter 12). This sparse group is a near-twin in size and brightness to H VII-4 but only a fraction as rich. This combo shows up well in a 6- to 8-inch telescope, but deciding just where one cluster ends and another begins is not easy! You will find them at the Taurus–Orion border, just over 1°NE of the star 15 Orionis. They can also be located by putting brilliant Aldebaran in a low-power eyepiece and sweeping about 8° due E of it.

Vulpecula

H VII-8 (NGC 6940): 20 35 + 28 18, open cluster, 6.3, 31′. "*Cluster, very bright, very large, very rich, considerably compressed, stars pretty large.*" Sitting all by itself just over 2°SE of 41 Cygni in a rich region of the Summer Milky Way is a lovely stellar jewel box that is largely ignored by observers. Here we find more than a hundred 8th-magnitude and fainter sapphires and diamonds, strikingly contrasted with a lone reddish jewel, spread out over an area about the size of the full Moon. That lone sun is the variable star FG Vulpeculae, which pulsates slowly from magnitude 9.0 to 9.5 over a period of 80 days. While even a 2-inch glass will show this big commune, it becomes ever more striking as aperture increases – being perhaps at its sparkling best as seen in an 8-inch scope using a low-power, wide-field eyepiece. It beckons us from across a distance of some 2,500 light-years (Fig. 9.3).

©·MDI 2005/6

Fig. 9.3. H VII-8 (NGC 6940) is a pretty but largely overlooked open cluster. Sitting in a rich area of the Summer Milky Way and containing more than a hundred stellar gems, it is definitely worth searching out. Courtesy of Mike Inglis.

Chapter 10

Showpieces of Class VIII

Coarsely Scattered Clusters of Stars

Listed below in alphabetical order by constellation are 14 of the most interesting objects in Herschel's Class VIII. Following the Herschel designation itself is the corresponding *NGC* number in parentheses, its Right Ascension and Declination (for Epoch 2000.0), the object's actual type (which may differ from the Class Herschel assigned it to), its visual magnitude, angular size in minutes (′) or seconds (″) of arc, and Messier or Caldwell number plus popular name if any. Next is a translation of Sir William's shorthand description (in italicized quotes) taken from the *NGC* itself, followed by comments from the author. These include directions for finding each object by sweeping for it, just as Herschel himself originally did.

Auriga

H VIII-71 (NGC 2281): 06 49 + 41 04, open cluster, 5.4, 15′. *"Cluster, pretty rich, very little compressed, stars pretty large."* This bright stellar commune is located just over 30′ S of the star χ −7 Aurigae, which lies in the same wide eyepiece field with it providing a nice contrast to the more distant cluster. The group contains several dozen members in an area about half the size of the full Moon. While easily visible even in a 3-inch glass, it perhaps looks its very sparkling best as seen in an 8-inch scope at low power.

Cassiopeia

H VIII-78 (NGC 225): 00 43 + 61 47, open cluster, 7.0, 12′. *"Cluster, large, little compressed, stars from the 9th to 10th magnitude."* Here is a nice loose cluster arranged roughly in the shape of the letter "W." Sweep for it about 2°NW of the star γ Cassiopeiae, in that constellation's familiar "W" (or "M") shape. Due to their sparseness, it is typically much easier to count the number of individual stars for objects in Herschel's Class VIII than in his richer, more compressed ones. In this case, about 20 of them can be seen in a 6-inch, set against a fairly rich Milky Way background. This is another of the discoveries of Caroline Herschel, which were generally made in the course of sweeping for comets.

Cygnus

H VIII-56 (NGC 6910): 20 23 + 40 47, open cluster, 6.7, 8'. *"Cluster, pretty bright, pretty small, poor, pretty compressed, stars from the 10th to 12th magnitude."* This is among the easiest of all the Herschel objects to find, since it sits right in the same eyepiece field as the prominent star γ Cygni at the center of the Northern Cross asterism. Lying just 30' to its NE, the cluster is neatly contrasted with the radiance of the star. But visually, this group may prove to be somewhat disappointing, depending on the size of telescope and magnification used. Despite the fact that some 40 members can be seen here in a 6- to 8-inch telescope within an area just 8' in extent, Herschel considered this cluster to be scattered and poor – as he did many of the objects in this Class. But at very low magnifications, it actually looks fairly rich. Apertures in the 3- to 4-inch range at low power provide an interesting view – that of a misty-looking little group of stars with a bright stellar jewel at its side set against a rich Milky Way field. And while here, take a moment to look up one of the most neglected of all the Messier objects – the little trapezoid-shaped cluster M 29, located just 1.5°S and slightly E of γ. It's been likened to a stubby-dipper or a tiny Pleiades cluster.

Lacerta

H VIII-75 (NGC 7243): 22 15 + 49 53, open cluster, 6.4, 21', = Caldwell 16. *"Cluster, large, poor, little compressed, stars very large."* This pretty cluster is only one of two objects in this Class that rated a spot in the Caldwell listing (the other being H VIII-20 in Vulpecula, below). Sweep for it about 2°W and just slightly S of α Lacertae, sitting in isolation in a beautiful Milky Way field. Here we find a coarse splash of some 40 massed star-jewels arrayed roughly in the shape of a triangle. Adding to the overall scene is a fairly dim but neat triple star located near the center of the cluster. This is Struve 2890, whose 8th-, 8th- and 9th-magnitude components are set 9″ and 73″ apart, respectively. Although H VIII-75 is visible in the smallest of telescopes, it benefits from aperture (combined with low-power and a wide field). Herschel regarded this as a poor object as seen in his big reflectors, but they had limited fields of view – especially at the relatively high magnifications he used in sweeping. The sight in a short-focus 10- to 12-inch Dobsonian reflector on a dark night is truly memorable! This scattered clan lies some 2,800 light-years from us.

Monoceros

H VIII-25 (NGC 2232): 06 27 – 04 45, open cluster, 3.9, 30'. *"Bright star (10 Monocerotis) plus cluster."* Now here is a Herschel object that is so bright that it can be seen with the unaided eye! But that is only because this cluster surrounds the 5th-magnitude star 10 Monocerotis. The dozen or so bright suns making up this group swarm around their blue-white central star like insects circling a streetlight at night. The individual members are bright enough to be seen even in a 2- or 3-inch glass but the view in a 6- to 8-inch is so much more satisfying. Yet, judging by Sir William's terse description, he apparently was not overly impressed with this object. After viewing H VIII-25, drop just slightly more than 2°S of it and take a look at the amazing star β Monocerotis. This is another of Herschel's discoveries – but in the realm of double and multiple stars (which, as followers in his footsteps, we should not ignore in our pursuit of his clusters and nebulae!). Better-known as Herschel's Wonder Star, this magnificent triple system is a sight never to be forgotten – whether using a 4-inch glass or a 14-inch one!

H VIII-5 = H V-27 (NGC 2264): 06 41 + 09 53, open cluster, 3.9, 20′, Christmas Tree Cluster. "*15 Monocerotis, cluster, double star, questionable nebulosity.*" This is another case of a cluster surrounding a naked-eye star, but this time there is also nebulosity involved. The star itself is 4th-magnitude 15 Monocerotis – which is also known as S Monocerotis, since it is slightly variable in brightness. The cluster is a big, bright gathering of some two dozen suns strikingly arranged in the shape of an upside-down Christmas tree! (Less imaginative observers see here instead an inverted arrowhead.) Obvious in the smallest of telescopes, it is perhaps best seen in a 6- to 8-inch instrument. 15 Monocerotis lies at the base of the tree and is the "double star" Herschel mentioned. It is actually a multiple sun – in addition to a 7.4-magnitude tight companion at just 3″ distance, there is also a 7.7-magnitude one at 156″, so this object is typically regarded as a triple system. However, there is also half a dozen other distant companions to as faint as 10th-magnitude, all of which are difficult to distinguish from the cluster stars themselves. The nebulosity that Herschel suspected (resulting in the dual designation given above) is what is known today as the Cone Nebula. It is a dark wedge or funnel of gas and dust extending S-ward from the top of the tree. While a striking spectacle as seen on photographs, visually it is elusive and difficult to glimpse except in large backyard telescopes under excellent sky conditions. Surprisingly, the beautiful Christmas Tree Cluster itself does not appear in the Caldwell listing. While it can be spotted directly due to the presence of its bright anchor star, it can also be picked up by sweeping about 6.5°S and just slightly E of much brighter γ Geminorum (Fig. 10.1).

Fig. 10.1. H VIII-5 = H V-27 (NGC 2264) is the Christmas Tree Cluster. This big, bright splash of stellar diamonds is arranged in the shape of an inverted tree, with the relatively bright star 15 Monocerotis marking the base of its "trunk." This close-up shows that star and part of the tree. Also glowing faintly beneath them is the Cone Nebula, which is H V-27 (thus this object's dual designation). Courtesy of Mike Inglis.

Ophiuchus

H VIII-72 (NGC 6633): 18 28 + 06 34, open cluster, 4.6, 27′. "*Cluster, little compressed, stars large.*" Despite this constellation's huge size – and even though it is simply loaded with globular clusters – this object is its only decent open cluster! It is yet another one of those wonderful finds of Caroline Herschel. Here we see a big, bright scattered clan of more than 50 stars – so bright, in fact, that it can actually be seen with the naked-eye object on a dark, transparent night. Its outline is somewhat elongated and asymmetrical, leading one observer of the past to describe it as "A lovely, great, straggling thing. . .of an absurd shape!" A wide field is key here rather than size of instrument – rich-field telescopes (RFTs) in the 4- to 8-inch aperture range show it beautifully. Sitting in relative isolation right on the Ophiuchus-Serpens border in a rich Milky Way starfield, it requires careful sweeping to find about 7°NW of the pretty double star θ Serpentis. Along the way you will encounter the big sparse open cluster IC 4756 in Serpens, which was missed by most of the early telescopic observers including Herschel due to its huge angular size – nearly twice that of the full Moon. (Incidentally, this is one of those very rare naked-eye objects actually having an *Index Catalog* number.) H VIII-72 itself lies some 1,000 light-years from us.

Orion

H VIII-24 (NGC 2169): 06 08 + 13 57, open cluster, 5.9, 7′, The "37" Cluster. "*Cluster, small, little rich, pretty much compressed, double star Struve 848.*" This little group of some 20 suns occupies the corner of a right triangle formed with the stars ξ and ν Orionis. Both of these sentries lie within 1° of the cluster and they can all be seen together in a rich-field telescope at lower power. This group seems rather close-packed for a Class VIII object and also looks bigger than the 7′ listed above. The arrangement of its stars has given rise to this cluster's unusual name. And indeed, with some imagination, the number 37 written in stars can actually be made out. Located within the cluster is the dim multiple star Struve 848. It consists of an 8th-magnitude primary, with 9th-, 9th- and 10th-magnitude companions having separations of 3″, 28″, and 43″, respectively. At least a 6-inch is needed to see them well. Note that Herschel saw it as just a double star (Fig. 10.2).

Puppis

H VIII-38 (NGC 2422): 07 37 – 14 30, open cluster, 4.4, 30′, = Messier 47. "*Cluster, bright, very large, pretty rich, stars large and small.*" This is another of the famed "missing" Messier objects resulting from an error in recording its position. It was found by Messier at the same time as the neighboring open cluster M46 containing the planetary H IV-39 (NGC 2438 – see Chapter 6) but was subsequently lost. For this reason, Herschel's discovery of this object was an independent one, explaining why M47 appears in his catalog. A line from β Canis Majoris extended

Fig. 10.2. H VIII-24 (NGC 2169) is known as the "37" Cluster from the distinctive arrangement of its stars. This object seems like it belongs in Class VII rather than in Class VIII since it is actually a fairly compact grouping. Courtesy of Mike Inglis.

through Sirius points right at M47. A leisurely sweep through this majestic area some 12°E and 2°N of Sirius itself brings you to it, with M46 just 1.5° to its E. This big scattered commune contains at least a dozen bright stars and many fainter ones set against the fine stardust of the Milky Way background itself. A number of lovely tinted gems can be seen amidst the glitter of its stellar sapphires and diamonds in 4- to 8-inch telescopes. There is also a neat double star near its center. This is Struve 1121, consisting of nearly matched 7.5-magnitude blue–white components set 7″ apart. Surprisingly, Herschel did not mention this pair despite the fact that it is quite obvious even in a 3-inch glass. H VIII-38/M47 lies at a distance of 1,500 light-years, amid a Winter wonderland of deep-sky treasures (Fig. 10.3).

H VIII-1 (NGC 2509): 08 01 – 19 04, open cluster, 9.3, 4′. "*Cluster, bright, pretty rich, little compressed, stars small.*" This little stellar jewel box is being included here because it does not seem to belong in Class VIII. Packing some 40 stars into just 4′ of sky, it appears rich and concentrated – not coarse and scattered! Officially, it is classified as a "very rich and compact cluster," which certainly better matches what is seen in the eyepiece. The fact that the stars in this group are rather dim, combined with its small size, means this object needs lots of aperture to be well seen – at least 10-inches or larger being recommended. It takes some careful star hopping and sweeping to find it. It lies nearly 6°NE of ρ Puppis, and about 2°W and slightly N of 16 Puppis, in a very rich Milky Way starfield. Having found it, what Herschel Class would you assign this object to?

Fig. 10.3. H VIII-38 is better-known today as M 47. But although this big, bright and rich stellar commune was cataloged by Messier, later observers could not find it due to an error in its reported position. Well before this mistake was eventually discovered, Herschel independently swept it up and included it in his catalog. Courtesy of Mike Inglis.

Scutum

H VIII-12 (NGC 6664): 18 37 – 08 13, open cluster, 7.8, 16′. *"Cluster, large, pretty rich, very little compressed."* This starry commune sits just 30′ E of the star α Scuti, which lies in the same eyepiece field – making this object a snap to find and providing a nice contrast with the much dimmer stars of the cluster itself. Seemingly little-known to observers, it is quite typical of the many worthy clusters, nebulae and galaxies overshadowed by the presence of a more famous and spectacular deep-sky wonder. In this case, it is the open cluster M11 to the NE of H VIII-12, better-known as Smyth's Wide Duck Cluster. Here we find some two dozen stars in a fairly large and scattered assemblage in another rich Milky Way field. As its member suns are on the dim side, at least a 6-inch scope on a dark night is required for a good view.

Taurus

H VIII-8 (NGC 1647): 04 46 + 19 04, open cluster, 6.4, 45′. *"Cluster, very large, stars large, scattered."* Located 3.5°NE of Aldebaran (α Tauri), this is another neglected wonder of the deep sky – in this case, one overshadowed by the glittering naked-eye Hyades Cluster just to its SW. Herschel's brief description does not really do justice to this lovely object. Here we find more than a dozen fairly bright stars along with many fainter ones spread out over an area larger than the full Moon. There are also a number of apparent star-pairs scattered about and two prominent suns on the S edge of the group. Officially, this cluster is listed as containing some 25 members of 8th-magnitude and fainter. And in a 3- or 4-inch glass at low power, this is pretty much what is seen. But in larger apertures it is a different story – providing they have a wide enough field of view to take in the entire cluster. In big short-focus Dobsonian reflectors, as many as 200 stars have been seen here by experienced observers. In that case, this would be considered a pretty rich open cluster – but Herschel did not indicate this in his account above. It is quite possible that at least some of these dim stars are part of the Milky Way background here and not actual members of the cluster itself.

Vulpecula

H VIII-20 (NGC 6885): 20 12 + 26 29, 6.7, 7′, = Caldwell 37. *"Cluster, very bright, very large, rich, little compressed, stars from the 6th to 11th magnitude."*/**H VIII-22 (NGC 6882): 20 12 + 26 33, 8.1, 18′.** *"Cluster, poor, little compressed."* Here is an interesting but virtually unknown double cluster of stars. H VIII-20 is the more obvious of the two and surrounds the naked-eye star 20 Vulpeculae. It is a fairly rich but scattered swarm of some three dozen suns. Its companion is H VIII-22, which overlaps it. The star 19 Vulpeculae sits on its N edge along with two other fainter ones, defining its boundary there, with 18 Vulpeculae lying just to the NW of them. This clan has less than half as many stars as H VIII-20 spread out over an area twice as large. As a result it appears quite sparse – so much so that Sir William called it a "poor" cluster. At least a 6-inch glass and a dark, transparent night are needed to really make out this twin stellar commune. Even then, it is a real challenge deciding just where one cluster ends and the other begins (or where either one is in the first place, if using smaller instruments). Incidentally, various sources list the visual magnitude of H VIII-20 as bright as 5th and as faint as 9th. These values are almost certainly incorrect. And the 5.5-magnitude given for H VIII-22 in the original Herschel Club's manual does not seem right either. The brightness's given above for these two clusters seem to match what actually greets the eye at the telescope. Also, note that the Caldwell list apparently recognizes only the more prominent member of this dual clan.

Chapter 11

Samples of Classes II & III

Faint Nebulae and Very Faint Nebulae

While the author has long promoted the merits of dropping the objects contained in Herschel's Class II (faint nebulae) and Class III (very faint nebulae) to arrive at a manageable target list for prospective Herschel Club members, it will be admitted that there are a few (a very few!) objects in those two classes that – while not showpieces in the normal sense of the word – are still fascinating telescopic targets for one reason or another. (While on the subject of celestial showpieces, the author has spent more than 50 years surveying the sky for its visual treasures with hundreds of different sizes, types and makes of telescopes – one fruit of which is the book *Celestial Harvest: 300-Plus Showpieces of the Heavens for Telescope Viewing and Contemplation*, which was reprinted by Dover Publications in 2002.)

In what follows, 18 (counting two dual designations) of the entries in Herschel's Classes II and III are listed in alphabetical order by constellation for readers wishing to examine them firsthand to get a feel for what objects in these two classes look like. Following the Herschel designation is the corresponding *NGC* number in parentheses, the Right Ascension and Declination (Epoch 2000.0), the object's actual type (which may differ from the Class Herschel assigned it to), its visual magnitude, angular size in minutes (′) or seconds (″) of arc, and Caldwell number plus popular name if any. Next is a translation of Sir William's shorthand description (in italicized quotes) taken from the *NGC* itself, followed by comments from the author. All of these targets are visible in an 8-inch aperture and many of them in telescopes half that size, given good sky conditions.

Andromeda

H II-224 (NGC 404): 01 10 + 35 37, galaxy, 10.1, 3′ × 3′, False Comet. "*Pretty bright, considerably large, round, gradually brighter in the middle, β Andr. south following.*" This small dim blur is situated just 6′ of arc NW of the star β Andromedae. So close it is to the star, in fact, that until relatively recently star atlases failed to plot it. As a result, many observers coming across it thought that they had discovered a comet. Others simply dismissed it as a reflection of the star itself. Fortunately, most good atlases today have a cut-out in the star-symbol for β so that the symbol for the galaxy can be shown as well. While this ghostly object can actually be glimpsed in a 3-inch glass, it takes at least a 6-inch to make its presence obvious.

Aquila

H III-743 (NGC 6781): 19 18 + 06 33, planetary nebula, 12.5, 105″, Soap Bubble Nebula. "*Planetary nebula, faint, large, round, very suddenly brighter in the middle of disc, small star north following.*" Although often pictured in astroimages, this dim sphere is quite elusive visually despite the fact that experienced observers claim it is within the reach of a 4-inch glass under dark skies. Some sources list it as a magnitude brighter than that given here – which still results in a very low surface brightness at the eyepiece (Fig. 11.1).

Cassiopeia

H II-707 (NGC 185): 00 39 + 48 20, galaxy, 9.2, 17′ × 14′, = Caldwell 18. "*Pretty bright, very large, irregularly round, very gradually much brighter in the middle, resolvable (mottled, not resolved).*" This is one of "the other two" companions to the Andromeda Galaxy (M31), the better-known and closer-in ones being M32 and M110. This object lies near the other companion, NGC 147, but as described in Chapter 12, Herschel did not see it. Both galaxies are roughly the same magnitude and apparent size, which – being fairly large – results in a low surface brightness. Note that he felt H II-707 was resolvable or mottled in appearance. Since they are

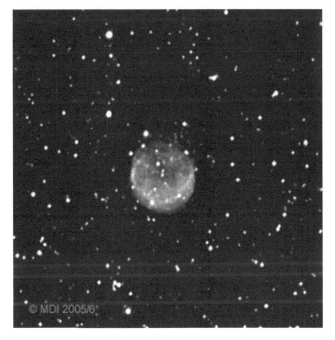

Fig. 11.1. Known as the Soap Bubble Nebula, H III-743 (NGC 6781) is quite faint visually in typical backyard telescopes. Sir William was certainly correct in placing it in his Class III. Courtesy of Mike Inglis.

located some distance from M31 itself (actually in a different constellation!), the author has always found it difficult to think of them as "companions" to the great Andromeda spiral.

Cetus

H II-6 (NGC 1055): 02 42 + 00 26, galaxy, 10.6, 8′ × 3′. *"Pretty faint, considerably large, irregular extended 80 degrees, brighter in the middle, a star of 11th magnitude north 1′."* A fairly dim edge-on spiral lying in the same eyepiece field 30′ to the NW of the well-known compact galaxy M77. Some sources list the Herschel object as faint as 12th-magnitude – which certainly better matches its appearance in small telescopes. The two objects are actually part of a small galaxy group, of which M77 is the brightest member (Fig. 11.2).

Coma Berenices

H II-391 (NGC 4889): 13 00 + 27 58, galaxy, 11.5, 3′ × 2′, = Caldwell 35. *"Pretty bright, pretty much extended, brighter in the middle, a star of 7th-magnitude north."* This diminutive object is the most visible member of the remote Coma Galaxy Cluster.

Some sources make it as faint as 13th-magnitude and less than half the size given here. In any case, it is certainly not an easy object in typical backyard telescopes. Were the author to guess its Herschel Class, it would certainly be assigned to III rather than II, so dim and elusive is it!

© MDI 2005/6

Fig. 11.2. H II-6 (NGC 1055) is a very dim spiral lying in the same wide eyepiece field with the bright galaxy M77. It appears much fainter in the eyepiece than its published magnitude would lead you to believe and most observers simply do not even realize that it is there! Courtesy of Mike Inglis.

Draco

H II-759 (NGC 5907): 15 16 + 56 19, galaxy, 10.4, 12′ × 2′, Splinter Galaxy.
"Considerably bright, very large, very much extended 155 degrees, very gradually, then pretty suddenly brighter in the middle to a nucleus." This is one of the flattest galaxies known, appearing as a long, thin, dim streak. As such, it is certainly well-named. While it can be glimpsed in 4- to 5-inch apertures on a dark night, it takes at least an 8- to 10-inch to appreciate the view. This object is actually one of the more fascinating sights to be found in Herschel's Class II and is considered a showpiece as seen in large-aperture Dobsonian reflectors (Fig. 11.3).

Gemini

H II-316/317 (NGC 2371/2372): 07 26 + 29 29, planetary nebula, 12.0, 55″. *"Bright, small, round, brighter in the middle to a nucleus, preceding of double nebula."/"Pretty bright, small, round, brighter in the middle to a nucleus, following of a double nebula."* This dim double-lobed planetary requires apertures of 12-inches or more and high magnifications on a steady night to see why it has two Herschel numbers. It is certainly a very easy object to miss when sweeping at low powers, as the author can attest from personal experience (Fig. 11.4).

© MDI 2005/6

Fig. 11.3. H II-759 (NGC 5907) is the well-named Splinter Galaxy. This rather dim spiral is best seen in medium- to large-aperture instruments, appearing as a very narrow ray or sliver of light. A really dark sky is needed to appreciate the view. Courtesy of Mike Inglis.

Fig. 11.4. H II-316/317 (NGC 2371/2372) is a dim bi-polar planetary rather than a nebulosity as Herschel classified it. But in all fairness to Sir William, it sure does not look like your typical planetary nebula. Courtesy of Mike Inglis.

Hercules

H II-701 (NGC 6207): 16 43 + 36 50, galaxy, 11.6, 3′ × 1′. "*Pretty bright, pretty large, extended 45 degrees +/–, very gradually much brighter in the middle.*" This tiny dim spiral would surely never have received notice from observers were it situated by itself. But it lies in the same wide eyepiece field with the magnificent Hercules Cluster (M13) – one of the largest, brightest, and finest globulars in the entire sky. Situated 30′NE just beyond the outer fringes of this glittering stellar beehive, it requires very careful attention to spot in 6-inch and smaller scopes. Here is one of the best examples of the extreme "depth of field" to be found in viewing deep-sky wonders, for the galaxy itself lies some 2,000 times further from us than M13's distance of 24,000 light-years (Fig. 11.5).

Hydra

II-196 (NGC 5694): 14 40 – 26 32, globular cluster, 10.2. 4′, = Caldwell 66. "*Considerably bright, considerably small, round, pretty suddenly brighter in the middle, resolvable (mottled, not resolved), a star of magnitude 9.5 south preceding.*" This dim little starball is somewhat elusive due to its small size and rather low Declination. It lies about 2°W and 30′S of the 5th-magnitude star 56 Hydrae. It definitely needs a dark night, steady skies, and at least an 8- to 10-inch aperture to make out its globular nature.

Fig. 11.5. H II-701 (NGC 6207) is a small, dim spiral galaxy lying just NE of the spectacular Hercules Cluster. Depending on sky conditions, this can be a tough object to spot even knowing its position in relation to the globular itself. It definitely belongs to Class II – and some observers after searching for it may even think it should be placed in Class III instead. Courtesy of Mike Inglis.

Leo

H II-44/45 (NGC 3190/3193): 10 18 + 21 50, galaxies, 11.0/10.9, 5′ × 2′/3′, Hickson Galaxy Group #44. *"Bright, pretty small, extended, pretty suddenly brighter in the middle to a nucleus."/"Bright, small, very little extended, pretty suddenly little brighter in the middle, star of 9.5-magnitude at 354 degrees, 80"."* These two small dim objects are the brightest members of a compact galaxy group located midway between the lovely double star Algieba (γ Leonis) and ζ Leonis in the Sickle asterism of Leo. Because they are situated in the midst of these two bright stellar landmarks, it is easy to be sure you are looking in the right place – but it is another matter to be sure you are actually seeing them in small instruments! At least a 6-inch aperture is recommended here. These two Herschel objects lie just a few minutes of arc apart in the eyepiece, floating in the midst of several other even fainter galaxies (Fig. 11.6).

Leo

H II-52 (NGC 3626): 11 20 + 18 21, galaxy, 11.0, 3′ × 2′, = Caldwell 40. *"Bright, small, very little extended, suddenly brighter in the middle."* This spiral is included here simply because Sir Patrick Moore found reason to place it in his Caldwell listing.

Fig. 11.6. H II-44/45 (NGC 3190/3193) are the two brightest members of a small knot of galaxies known as the Hickson Galaxy Group #44 located midway between the stars γ and ζ in the head of Leo. Here the usage of the word "brightest" may be deceiving, for these dim objects surely deserve being placed by Herschel in his Class II. Courtesy of Mike Inglis.

However, it is just another small, dim, and visually unimpressive galaxy – quite typical of the multitude of such objects populating Herschel's Classes II and III.

Pegasus

H II-240 (NGC 7814): 00 03 + 16 09, galaxy, 10.6, 6′ × 2′, = Caldwell 43/Electric Arc Galaxy. *"Considerably bright, considerably large, extended, very gradually brighter in the middle."* This edge-on spiral gets its name from its striking appearance on photographs, where its equatorial dust lane and nuclear bulge do cause it to look like its namesake. But visually, it takes at least a 12- to 14-inch aperture to see the effect. Some sources make this object as faint as 12th-magnitude, which does better match its aspect in the eyepiece (Fig. 11.7).

Sagittarius

H II-586 (NGC 6445): 17 49 – 20 01, planetary nebula, 12.0, 38″ × 29″, Crescent Nebula. *"Pretty bright, pretty small, round, gradually brighter in the middle, resolvable (mottled, not resolved), star of magnitude 15 north preceding."* This challenging object is one of two nebulosities bearing the name Crescent Nebula, the other being H IV-72 (NGC 6888) in Cygnus, described in Chapter 6. Some observers rate the brightness here as at least a magnitude fainter, for this is indeed a dim object – one that could easily have been put in Herschel's Class III. Note that he thought it was resolvable or mottled in appearance. The only real interest here is that the globular cluster H I-150 (NGC 6440) lies in the same eyepiece field (as described in Chapter 5), offering quite a striking contrast between the two classes!

Fig. 11.7. H II-240 (NGC 7814) is known as the Electric Arc Galaxy from its striking photographic appearance. Visually it is quite dim and seems at least a magnitude fainter than its published value. Courtesy of Mike Inglis.

Triangulum

H III-150 (NGC 604): 01 34 + 30 39, diffuse nebula, 11.0, 1′. "*Bright, very small, round, very very little brighter in the middle.*" What makes this tiny dim object noteworthy as a nebula is that it lies in another galaxy! It is embedded in one of the graceful spiral arms of the Triangulum Galaxy (sometimes also called the Pinwheel Galaxy, even though the same name is given to M101 in Ursa Major). Located about 10′ of arc NE of the galaxy's nucleus, this little cloud is a pocket of star formation known as an H-II region (the designation having no relation to Herschel) some 30 times the size of the famed Orion Nebula (M42/M43) in our Galaxy. This accounts for its visibility across such a vast distance – some 3,000,000 light-years. While a 4-inch glass will show it, twice that aperture is needed to see it well (Fig. 11.8).

Virgo

H II-75 (NGC 4762): 12 53 + 11 14, galaxy, 10.2, 9′ × 2′, The Kite. "*Pretty bright, very much extended 31 degrees, three bright stars south, following of two [nebulae].*" This edge-on spiral has an unusual extension at one end reminding early observers of a celestial kite, and is neatly grouped with three stars. Adding to the scene here is the 10.6-magnitude 5′ × 3′ barred spiral H I-25 (NGC 4754) lying just 11′ to the SE. This presents a great opportunity to visually compare objects of Classes I and II very close together in the eyepiece, at least a 6-inch glass being recommended for doing so. (Note that although H II-75 is listed as brighter than H I-25 by nearly half a magnitude, the former *appears* fainter than the latter due to its larger size and resulting lower surface brightness.)

Fig. 11.8. H VIII-150 (NGC 604) is one of the several bright knots of nebulosity embedded within the huge spiral galaxy M33, better-known as the Triangulum or Pinwheel Galaxy. It is one of the very few objects that can actually be glimpsed *within another galaxy* in small telescopes across the vastness of intergalactic space. (H VIII-150 itself actually lies just outside the bottom left edge of this image, which encompasses the complex central portion of the galaxy.) And here is a case where some observers may feel this object should be upgraded to Herschel's Class II since it does not seem as difficult to see as his typical Class III entries. Courtesy of Mike Inglis.

Virgo

H II-297 (NGC 5247): 13 38–17 53, galaxy, 10.5, 5′. "*Remarkable, very much so, considerably faint, very large, very gradually, then pretty suddenly much brighter in the middle to a large nucleus.*" There are some puzzles involving this object and Herschel's description of it in the *NGC*. He was obviously very excited about it, giving it "two thumbs up" (in his shorthand notation), meaning that he considered it as "remarkable – very much so." But then he goes on to describe it as "considerably faint" and "very large" – neither of which comments match up with the current catalog data given above. Could he possibly have been looking at another object and mis-cataloged it as H II-297? If so, what and where is it? In any case, if the magnitude of this galaxy is really 10.5 as listed, it certainly does not belong in Class II. But some sources do consider it to be as faint as 12th-magnitude, better fitting what Herschel saw and how he assigned it.

Chapter 12

Showpieces Missed by Herschel

How Did This Happen?

As diligent and thorough an observer as he was, William Herschel passed over a number of clusters, nebulae, and galaxies that are routinely viewed by deep-sky observers today using typical backyard telescopes. In some cases this is surely understandable – as for those objects apparently too small or too large in angular size for the telescopes he was using. In other cases, it is a real mystery how he ever missed them. Perhaps changing atmospheric conditions, the cooling of the metal mirrors and their varying reflectivity as they tarnished, and fatigue of the observer are among the explanations for these. There is also the problem of sweeping near the zenith using altazimuth-mounted telescopes as were Herschel's, which create "dead spots" in the sky. But surely he would have allowed for this. (And speaking about observing near the zenith, we can only imagine what it was like attempting to look into the mouth of the vertically pointing tube when used as "front-view" as Herschel did. In the case of his mammoth 40-foot reflector, this placed him more than 40 feet above the ground, the supporting structure itself towering some 50 feet high!)

Initially his sweeps were made by moving the telescope in azimuth, but this was then changed for vertical sweeps in which the telescope was raised or lowered in altitude. In both cases, the eyepiece field in the sweep just completed overlapped that of the next one – much the same technique that visual comet hunters still employ today in sweeping – so that no part of the sky would be overlooked between sweeps. There is no question that Herschel not only scrutinized the entire heavens visible from his latitude (his southernmost discovery lying low above the horizon at a Declination of about −33°, as mentioned in Chapter 3), but he did so *several times* in his various reviews.

List of Overlooked Showpieces

In what follows, 22 of the objects missed by Sir William are listed in alphabetical order by constellation for readers wishing to examine them firsthand and perhaps come to their own conclusions as to why they were overlooked. Following the *NGC* or *IC* designation are the Right Ascension and Declination (Epoch 2000.0), the type of object, its visual magnitude and angular size in minutes (′) or seconds (″)

of arc, and Caldwell number plus popular name if any. A brief commentary follows on why it may have been overlooked. All of these targets are visible on a dark night in 4- to 8-inch telescopes and a few even in binoculars.

Aquarius

NGC 7293: 22 30 – 20 48, planetary nebula, 6.5, 16′ × 12′, = Caldwell 63/Helix Nebula. While this is the brightest object of its class in the sky, it is also huge in apparent size, resulting in a very low surface brightness in the eyepiece. But Herschel found many other big and faint nebulous objects. Needs a dark, transparent night without any moonlight (Fig. 12.1).

Aquila

NGC 6709: 18 52 + 10 21, open cluster, 6.7, 13′. The best of Aquila's clusters, with some 40 member-stars in a scattered grouping. An easy object in all apertures – and no obvious reason why Herschel would not have seen it!

Auriga

IC 405: 05 16 + 34 16, diffuse nebula, 9.2, 30′ × 20′, = Caldwell 31/Flaming Star Nebula. Faintish nebulosity surrounding the variable star AE Aurigae. This object should have easily been within the reach of Herschel's larger telescopes.

Fig. 12.1. The Helix Nebula (NGC 7293) was overlooked by Herschel even though it is the biggest and brightest planetary in the sky. But in fairness to Sir William, its huge size not only requires a large field of view (which his reflectors did not have) to encompass it but it also results in an extremely low surface brightness. Photographed using an 11-inch Schmidt-Cassegrain telescope. Courtesy of Steve Peters.

Camelopardalis

IC 342: 03 47 + 68 08, galaxy, 9?, 16′, = Caldwell 5. This object has highly discordant published magnitudes ranging from 8th to 12th. It is certainly not as bright as the 8.4 given in the Caldwell list (if it were, it would be one of the brightest in the sky!) and if it is as faint as 12th it could have conceivably been overlooked by Herschel. But here again, he found many others even fainter.

Cassiopeia

NGC 147: 00 32 + 48 30, galaxy, 9.5, 18′ × 11′, = Caldwell 17. One of "the other two" companions of the Andromeda Galaxy (M31), the better-known and closer-in ones being M32 and M110. This object and H II-702 (NGC 185 – see Chapter 11) lie quite some distance from M31 in the sky (but very close to each other) and both are fairly bright for such extended big galaxies. While Herschel did see 9.2-magnitude H II-702, he missed neighboring NGC 147.

NGC 281: 00 53 + 56 36, diffuse nebula, 7.8, 35′ × 30′. This fairly obvious fuzzy patch has a small clump of stars embedded in it. Both the stars and nebulosity have been spotted in 25 × 100 binoculars, and they should have been unmistakable in Herschel's telescopes (Fig. 12.2).

Fig. 12.2. NGC 281 is a relatively bright diffuse nebulosity encompassing several dim stars. Visible even in small telescopes, it is really a bit of a mystery how Herschel overlooked it in his extensive sweeps of the sky. Courtesy of Mike Inglis.

Cepheus

NGC 188: 00 44 + 85 20, open cluster, 8.1, 14′, = Caldwell 1/The Ancient One. This dim cluster is famous as being the oldest object of its kind known (age at least 12 *billion* years!). While not spectacular in any sense of the word in amateur instruments, it still should have been obvious to Herschel. While sweeping at such high declinations (in this case less than 5° from the North Celestial Pole) might be a problem for a big equatorially mounted telescope, it should not be for altazimuths like those that Sir William exclusively used.

Cetus

IC 1613: 01 05 + 02 07, galaxy, 9.2, 19′ × 17′, = Caldwell 51. Despite its magnitude, this object is rather faint due to its large angular extent. Yet it is not difficult to see in big Dobsonian reflectors, which are equivalent in performance to Herschel's 20-foot telescopes.

Cygnus

NGC 6819: 19 41 + 40 11, open cluster, 7.3, 5′, Foxhead Cluster. This small V-shaped clan contains some 150 members afloat in the Summer Milky Way. Perhaps its diminutive size combined with the rich stellar background caused it to be passed over – something the author himself has done in sweeping for it. Although actually seen by Caroline Herschel, for some unknown reason it was not included in the Herschel catalog (Fig. 12.3).

Fig. 12.3. NGC 6819, better known as the Foxhead Cluster, is a small compact and rich group of stars embedded within the Summer Milky Way. Overlooked by Sir William, it *was* seen by his sister Caroline yet not placed in the Herschel catalog – despite the fact that most of her other finds were included. Courtesy of Mike Inglis.

NGC 7027: 21 07 + 42 14, planetary nebula, 9.3, 18″ × 11″, Stephen's/Webb's Protoplanetary. This eerie-looking, intensely blue egg is only about half the apparent size of its famous neighboring planetary in Cygnus H IV-73 (NGC 6826) which Herschel did discover, probably accounting for why he missed it. The latter is better-known as the Blinking Planetary (see Chapter 6) and like it, the former "blinks" as well!

IC 5146: 21 53 + 47 16, diffuse nebula, 9.3, 10′, = Caldwell 19/Cocoon Nebula. This circular patch is immersed in the Summer Milky Way and is easy to overlook – not due to its size in this case but rather its dimness. But Herschel found many just as inconspicuous as this one.

Fornax

NGC 1360: 03 33 – 25 51, planetary nebula, 9.4, 6′ × 4′. A big dim disk of light with an 11th-magnitude central star. It has been seen in a 2-inch glass and is definite in a 6-inch. But its low surface brightness (resulting from its huge size) combined with its low altitude as seen from England are likely responsible for its being passed over.

Hercules

NGC 6210: 16 44 + 23 49, planetary nebula, 9.3, 20″ × 16″, = Struve 5N. This tiny but intense bluish disk can be seen in a 3-inch telescope with careful attention. Visually it does seem much smaller than indicated here. Even so, Herschel discovered Uranus which is just over 4″ in apparent size at its largest (although it is several magnitudes brighter than this object). He also similarly missed NGC 6572/Struve 6N in Ophiuchus (see below) – a near twin of this little jewel. But, so too, did all the other early nebula hunters! It fell to the great double star observer Wilhelm Struve to find these two small planetaries during the course of his double star surveys with an optically superb 9.6-inch long-focus (f/18) refractor (Fig. 12.4).

Lyra

NGC 6791: 19 21 + 37 51, open cluster, 9.5, 16′. This big dim but extremely rich swarm contains over 300 stars. It is quite easy to pass right over it in sweeping, but once seen it can be glimpsed in a 5-inch scope on a dark night and is a fascinating sight in an 8-inch It looks like a fainter version of H VI-30 (NGC 7789 – see Chapter 8) in Cassiopeia, which is a showpiece of the Herschel catalog discovered by William's sister Caroline. He would surely have loved adding this gem to his list (Fig. 12.5).

Monoceros

NGC 2237: 06 32 + 05 03, diffuse nebula, —, 80′ × 60′, = Caldwell 49/Rosette Nebula. This huge ring of nebulosity apparently was not seen by Herschel even though he did discover the cluster of stars H VII-2 (NGC 2244 – see Chapter 9) within it! Like

Fig. 12.4. NGC 6210/ Struve 5N is a bright little planetary nebula that looks nonstellar even at low powers, while medium magnifications clearly reveal its bluish disk. Yet it was strangely missed by all the early observers including William Herschel himself. Courtesy of Mike Inglis.

Fig. 12.5. NGC 6791 is a big, very rich open cluster that is little-known to observers due to its dimness. It looks more like a very loose globular cluster than an open one as seen in large amateur instruments. Courtesy of Mike Inglis.

Fig. 12.6. The large ring-shaped Rosette Nebula (NGC 2237) was apparently not seen by Herschel even though he did discover the star cluster within it (H VII-2/NGC 2244). The limited fields of view of his big reflectors may account for this, although in sweeping he must surely have passed right over various parts of the ephemeral ring itself, only a portion of which is shown here. Courtesy of Mike Inglis.

many others then and now, he may have been looking right through the central hole without realizing it, for this object is twice the apparent size of the full Moon! However, sections of the nebulosity are definitely bright enough to have been detected in his telescopes. The entire Rosette Nebula and associated Rosette Cluster can be seen in binoculars and rich-field telescopes (RFTs) (Fig. 12.6).

Ophiuchus

IC 4665: 17 46 + 05 43, open cluster, 4.2, 41′, Summer Beehive. This clan of more than 30 suns is quite bright, but it is also very large and scattered. While striking in binoculars and RFTs, it is way too big for typical telescopic fields – explaining why it was long overlooked by observers and does not appear in the *NGC*. Strangely, however, Caroline Herschel *did* see it and showed it to her brother! But for some unknown reason, it never made it into the Herschel catalog.

NGC 6572: 18 12 + 06 51, planetary nebula, 9.0, 15″ × 12″, = Struve 6N. This small intense bluish sphere was missed by Herschel and others for the same reason given for its near-twin NGC 6210 in Hercules, as discussed above. Here again, this object seems smaller in the eyepiece than the published angular size suggests yet it can be seen in a 3- or 4-inch glass.

Orion

NGC 1981: 05 35 – 04 26, open cluster, 4.6, 25′. This scattered group of a dozen or so stars lies just above a faint nebulous complex. Strangely, although he could not have missed seeing the cluster, Herschel did not catalog it. But he did log the

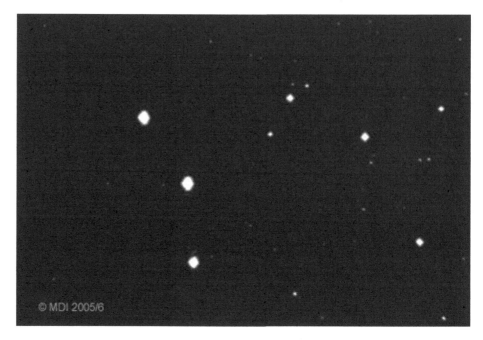

Fig. 12.7. NGC 1981 is a large and very loose handful of stars lying N of the Orion Nebula (M42/M43). Most observers (including Herschel himself) bypass this object as being simply a collection of field stars. The view in a rich-field telescope (RFT), however, does show it as the sparse open cluster it actually is. Courtesy of Mike Inglis.

nearby nebulosity, giving it his designation H V-30 (*NGC* 1977 – see Chapter 7). As an aside here, the famous observer William H. Pickering included NGC 1981 in his often-quoted list of the 60 finest objects in the sky – something the author has always questioned, for this surely is no showpiece! Apparently Herschel agreed and did not even bother to recognize it as a cluster (Fig. 12.7).

Sagittarius

NGC 6822: 19 45 – 14 48, galaxy, 8.8, 16′ × 14′, = Caldwell 57/Barnard's Dwarf Galaxy. At magnitude 8.8, this object might seem to be an easy target. But although it was discovered visually with a 5-inch refractor by the eagle-eyed Barnard, it is definitely not! Part of the reason is its extremely low surface brightness resulting from its great apparent size. But more of a factor is that visual magnitudes as faint as 11.0 have been assigned to it by various observers! What is really surprising here is that Herschel discovered the planetary H IV-51 (NGC 6818 – see Chapter 6) which lies just 45′ NNW of the galaxy but did not notice it. But given the small fields of view of his various telescopes, perhaps this is not so unexpected after all.

NGC 6530: 18 04 – 24 20, open cluster, 4.6, 15′. This sparse loose cluster of stars appears to be projected against the eastern side of the big, bright Lagoon Nebula (M8) and is unmistakable even in the smallest of telescopes. Why Herschel did not

Fig. 12.8. NGC 6530 is the open cluster associated with the Lagoon Nebula (M8), seen at bottom. It is obvious in the smallest of telescopes, yet Herschel did not catalog it for some unknown reason. The smaller, fainter Trifid Nebula (M20) appears at top. Photographed using an 8-inch Newtonian reflector. Courtesy of Steve Peters.

catalog it is a mystery – unless, of course, he simply considered it to be part of the Lagoon itself (which in actuality it is, for its suns were spawned by the nebula!) and not a separate object in its own right (Fig. 12.8).

Taurus

NGC 1554/5: 04 22 + 19 32, diffuse nebula, —, 30″, Hind's Variable Nebula. This tiny patch of nebulosity surrounds the erratic variable star T Tauri, which shines unsteadily at around 9th-magnitude. The nebula varies in visibility with the brightness of the star itself. While it can be seen at high power in a 4-inch glass, this is only after knowing of its existence. To discover something so small and dim would have taken a much larger telescope (as per one of Sir William's dictums, discussed in Chapter 4). Given the minuteness of this object and the fact that it may have been at minimum brightness at the time he was sweeping here, it is perhaps not surprising that Herschel missed it.

NGC 1807: 05 11 + 16 32, open cluster, 7.0, 17′. This sparse group of some 20 stars was not cataloged by Herschel – despite the fact that it lies almost on top of another cluster in the same eyepiece field which he did log! That object is H VII-4 (NGC 1817 – see Chapter 9), a near-twin in size and brightness but several times as rich as NGC 1807. This dual stellar clan shows up well in a 6-inch glass, but deciding just where one cluster ends and the other begins is not at all easy.

The "Missing" Herschel Objects

Where Did They Go?

In a situation analogous to the saga of the famed "missing" Messier objects (which have now all been accounted for as errors in identification and/or position), there is the case of objects that Herschel discovered and cataloged but which reportedly cannot be found in the sky today! These "disappearances" have mostly involved entries in his Class VIII, which are coarsely scattered clusters of stars – many of which were described as "poor" by him. As such, they can often be difficult to pick out from the stellar background since most open clusters lie along the plane of the Milky Way's rich stratum of stars.

The modern story of Herschel objects apparently having vanished from the sky really dates back to 1973 when *The Revised New General Catalogue of Nonstellar Astronomical Objects* (or *RNGC*) was published by astronomers Jack Sulentic and William Tifft. The primary reference for this comprehensive work was the photographs taken for the famed National Geographical Society-Palomar Observatory Sky Survey with the 48-inch Schmidt camera on Palomar Mountain, home of the 200-inch Hale reflector. Objects that could not be identified on the large-scale plate prints were given a "type" code of "7" in the *RNGC*, meaning they are "nonexistent." These included no fewer than 30 of the 88 clusters in Herschel's Class VIII alone! (See the listing of these objects below.) As a point of interest, it should be mentioned here that the *RNGC* compilers also rejected a number of John Herschel's clusters as not existing, in addition to those of his father.

In 1975 two Canadian amateur astronomers – Patrick Brennan and David Ambrosi – began examining the sky for these missing objects using a 6-inch reflector. As Brennan later wrote, "Have you ever encountered a 'nonexistent' *RNGC* cluster alive and well, so to speak?" These observers found that many of the rejected Herschel clusters actually *were* visible in the eyepiece even if they were not distinguishable on the Palomar prints and so were not really missing after all. But as we are about to see, there is at least one case where a Herschel object *has* apparently vanished from the sky!

The Disappearance of H VIII-44

On all the editions of the widely used *Norton's Star Atlas* up through the 17th one, you will find the open cluster H VIII-44 (NGC 2394) plotted near the bright star Procyon in Canis Minor. Sitting as it does right on top of the star η Canis Minoris – between Procyon and β Canis Minoris – its position is a snap to locate. Sir William described this object as a large poor cluster containing bright stars. Yet it is nowhere to be found! The author has repeatedly looked for it using telescopes from 3- to 30-inches in aperture under all sorts of sky conditions without any success. The late Walter Scott Houston, one of the greatest visual observers ever and writer of *Sky & Telescope*'s "Deep-Sky Wonders" column for nearly half a century, had a similar experience. He reported that "The only problem is that the cluster is not there. I have searched for it several times with 4- to 16-inch telescopes. Since its cataloged position places it in the same low-power field as Eta, there is no question that I was looking in the right place." Why not explore this area of the sky for yourself and see if you can find this missing object? (Its coordinates and Herschel's actual description appear in the listing below.) (Fig. 13.1)

Some "Nonexistent" Herschel Objects

The following table contains the 30 objects in Herschel's Class VIII that are listed in the *RNGC* as nonexistent. It is being offered for those readers who may want to follow in the Canadian observers' footsteps and see for themselves if these clusters are truly missing from the sky. (More "nonexistent" objects can be found in his other classes as well – see those with "NE" in the type column in Appendix 3.) Arranged alphabetically by constellation, it gives the Herschel designation, the corresponding *NGC* number in parentheses, and the Right Ascension and Declination for Epoch 2000.0. Centering your telescope at the position indicated, you can perform a Herschelian "sweep" on this spot in the sky – going back and forth (or up and down), overlapping each sweep by part of the eyepiece field as you do. This technique will hopefully bring you the thrill of "discovering" the object yourself just as Herschel did originally!

It would also certainly be most helpful to know the visual magnitudes and apparent sizes of the objects you are searching for, but since they are considered

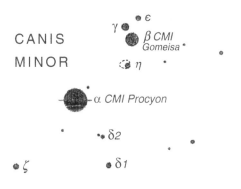

Fig. 13.1. The starfield surrounding Procyon and Gomeisa (α and β Canis Minoris, respectively). One of the most notorious missing Herschel objects is H VIII-44 (NGC 2394). Sir William described it as a large, bright but poor cluster, and his position puts it in the same eyepiece field with, and nearly on top of, the star η as shown by the dashed circle in this sketch. However, it is nowhere to be found.

nonexistent there is no catalog data available for them! Instead, a translation of Sir William's shorthand description (in italicized quotes) taken from the *NGC* itself is provided as an aid in identifying them. The frequent appearance of the words "poor" and "very little compressed" go a long way towards explaining why these objects have been so elusive to observers. Incidentally, out of the more than 2,700 nonstellar deep-sky objects listed in *Sky Catalogue 2000.0* (the database for the magnificent *Sky Atlas 2000*.0), there is only a *single* Herschel object indicated as nonexistent that also happens to have been designated as such in the *RNGC* – the open cluster H VIII-14 (NGC 6647) in Sagittarius. The actual notation under its listing of open clusters reads "No cluster." However, the *NGC 2000.0 does* show this as an existing object, giving a photographic magnitude of 8 but no apparent angular size. So it is being included both in the list below and the table in Appendix 3, where the type there is noted as "OC/NE?" with a reference to this chapter. Finally, the big gap in *NGC* numbers from 2678 to 6561 is the result of these clusters being predominantly located in the Winter and Summer Milky Way of the Northern Hemisphere sky. Good hunting!

Aquila

H VIII-13 (NGC 6728), 19 00 – 08 57: "*Cluster, very large, poor.*"

Aquila

H VIII-73 (NGC 6828), 19 50 + 07 55: "*Cluster, poor, little compressed.*"

Auriga

H VIII-49 (NGC 2240), 06 33 + 35 12: "*Cluster, pretty large, poor, very little compressed, stars of magnitude 7 and 10 to 15th.*"

Cancer

H VIII-10 (NGC 2678), 08 50 + 11 20: "*Cluster, very little compressed, poor.*"

Canis Major

H VIII-45 (NGC 2358), 07 17 – 17 03: "*Cluster, poor, little compressed.*"

Canis Minor

H VIII-44 (NGC 2394), 07 29 + 07 02: "*Cluster, large, poor, very little compressed, stars large [bright].*"

Cepheus

H VIII-63 (NGC 7234), 22 12 + 56 58: "*Cluster, small, poor, little compressed.*"

Cepheus

H VIII-62 (NGC 7708), 23 34 + 72 55: "*Cluster, large, poor, little compressed, stars of magnitude 8 and 10 to 15.*"

Cetus

H VIII-29 (NGC 7826), 00 05 – 20 44: "*Cluster, very poor, very little compressed.*"

Cygnus

H VIII-86 (NGC 6874), 20 08 + 38 14: "*Cluster, poor, little compressed.*"

Cygnus

H VIII-83 (NGC 6895), 20 16 + 50 14: "*Cluster, pretty rich, little compressed.*"

Cygnus

H VIII-82 (NGC 6989), 20 54 + 45 17: "*Cluster, considerably large, stars pretty small [faint].*"

Cygnus

H VIII-57 (NGC 7024), 21 06 + 41 30: "*Cluster, poor, little compressed, stars 10th magnitude and fainter.*"

Delphinus

H VIII-23 (NGC 6950), 20 41 + 16 38: *"Cluster, poor, very little compressed."*

Gemini

H VIII-9 (NGC 2234), 06 29 + 16 41: *"Cluster, extremely large, pretty rich, little compressed, stars large [bright] and small [faint]."*

Monoceros

H VIII-48 (NGC 2260), 06 38 − 01 28: *"Cluster, very large, poor, very little compressed, stars large [bright] and small [faint]."*

Monoceros

H VIII-51 (NGC 2306), 06 55 − 07 11: *"Cluster, poor, very little compressed."*

Monoceros

H VIII-1B (NGC 2319), 07 01 + 03 04: *"Cluster of very scattered stars of magnitude 8 to 9 and fainter."*

Orion

H VIII-2 (NGC 2063), 05 47 + 08 48: *"Cluster, poor, small, scattered stars."*

Orion

H VIII-6 (NGC 2180), 06 10 + 04 43: *"Cluster, pretty rich, little compressed, stars large [bright] and small [faint]."*

Puppis

H VIII-52 (NGC 2413), 07 33 − 13 06: *"Cluster, very large, poor, very little compressed."*

Puppis

H VIII-47 (NGC 2428), 07 39 – 16 31: "*Cluster, very large, very little compressed.*"

Puppis

H VIII-46 (NGC 2430), 07 39 – 16 21: "*Cluster, very large, very little compressed.*"

Sagittarius

H VIII-54 (NGC 6561), 18 10 – 16 48: "*Cluster, large, little compressed, stars considerably large [bright].*"

Sagittarius

H VIII-14 (NGC 6647), 18 32 – 17 21: "*Cluster, large, rich, little compressed, stars very small [faint].*"

Taurus

H VIII-41 (NGC 1802), 05 10 + 24 06: "*Cluster, stars considerably scattered.*"

Taurus

H VIII-4 (NGC 1896), 05 25 + 20 10: "*Cluster, very large, rich, very little compressed, stars from magnitude 9 to 12.*"

Taurus

H VIII-42 (NGC 1996), 05 38 + 25 49: "*Cluster, large, little compressed, little rich.*"

Taurus

H VIII-28 (NGC 2026), 05 43 + 20 07: "*Cluster, little rich, little compressed, stars pretty large [bright].*"

Vulpecula

H VIII-17 (NGC 6938), 20 35 + 22 15: *"Cluster, very large, poor, and very little compressed."*

Chapter 14

Conclusion

Herschel's Legacy

As stated in the opening to this work, Sir William Herschel was without question the greatest visual observer ever, as well as the greatest telescope-maker of his time. The fact that he accomplished so much in the field of astronomy without any formal education – using instruments made entirely by his own hands – is truly an inspiration for all of us who love the stars. The vows made by this unknown "amateur" to "carry improvements in telescopes to their utmost extent" and "to leave no spot of the heavens unvisited" were unprecedented in the long annals of astronomy. Truly, here was a man fired with a visionary passion for the heavens!

Those who knew Herschel intimately characterized him as "a man without a wish that has its object in the terrestrial globe," claiming that the King had "not a happier subject" and that "his expectations of future discoveries are so sanguine as to make his present existence a state of almost perfect enjoyment." You who have not only read this book but also gone further and actually seen with your own eyes many of Herschel's discoveries must surely have sensed what an incredibly heady and exciting time this was for him. In this day of intense specialization by professional (and many amateur) astronomers, it is often lamented that the fun and soul have gone out of observing the night sky. For many, it is the aesthetic and philosophical (and, for some, spiritual) aspects of astronomy that constitute its greatest value and charm. Stargazers at all levels now more than ever need to heed the dual legacy left by this truly amazing man – *to approach the heavens with a sense of wonder and awe, as well as with scientific curiosity and scrutiny*. This anticipation of encountering the wondrous unknown has perhaps never been better expressed than in these poetic lines by Charles Edward Barns in his long out-of-print-classic *1001 Celestial Wonders*:

> Lo, the Star-lords are assembling,
> And the banquet-board is set:
> We approach with fear and trembling,
> But we leave them with regret.

"The Construction of the Heavens"

William Herschel wrote that "A knowledge of the construction of the heavens has always been the ultimate object of my observations." What a bold and audacious concept this was! Prior to him, astronomers were concerned only with solar system

Fig. 14.1. William Herschel in the twilight of his later years. Here he looks contemplative and a bit disillusioned – perhaps over the performance of the great 40-foot telescope which, sadly, for the most part failed to meet his lofty expectations. Courtesy of the Royal Astronomical Society/Science Photo Library, London.

and fundamental positional astronomy, where the stars were mainly used as reference points for the positions of the Moon and planets, as well as for time-keeping and navigation. No one had thought about what might lie beyond them – or even (with a very few exceptions) knew that there were such objects as clusters, nebulae, and galaxies out there in deep space. While Charles Messier initially opened this frontier to some extent with his famous but short list of "comet imposters," it was Herschel who really "broke through the barriers of the skies" (as inscribed on his headstone*) and threw open the doors to the modern telescopic exploration of the universe (Fig. 14.1).

Some appreciation of Herschel's immense scientific legacy can be gained by the following examples of his work – all of which were in addition to his many discoveries in both the solar system and beyond. In his goal to study the structure of the heavens, he performed statistical star-counts over various parts of the sky – a process referred to as "gauging" the universe (his actual spelling was "gaging"). In connection with this, he calculated the "space penetrating power" (his term) of his

---

*Herschel made his last observation on June 1, 1821, and died on August 25, 1822, at the age of 84. He is interred in the little church at Upton, England, where he was married and near which the tube of his great 40-foot telescope was made. The actual words on his headstone read in Latin: "Coelorum perrupit claustra" ("He Broke Through the Barriers of the Skies.")

various telescopes, fully realizing that the larger the aperture the deeper into space a given instrument could look. These studies showed beyond any doubt that the Milky Way in which we live is, indeed, a flattened disk of countless suns.

In 1783, from the proper motions of just 13 stars, Herschel also discovered that our Solar System is moving through space in the direction of the constellation Hercules – just 10° from the modern position of the "Apex of the Sun's Way" as it is known in neighboring Lyra. He noticed that the stars in that direction of the sky appeared to be "opening up" or spreading out before us while those in the opposite part of the sky were "closing in" on themselves, much as the scenery in a moving car does looking out the front and rear windows, respectively. This was truly an amazing deduction on his part, especially given the relatively primitive and limited number of positions available to him.

Perhaps Herschel's most important single discovery was that of binary motion in double stars. At that time, such pairings were considered to be merely chance alignments of two unrelated objects lying at different distances (known as "optical doubles" today). Indeed, he employed this belief in an attempt to measure stellar parallax – the periodic shifting of closer stars with respect to more distant ones resulting from the orbital motion of the Earth about the Sun, from which it becomes possible to calculate distances trigonometrically. He presented a paper to the Royal Society on parallaxes in 1781. But in this attempt he failed, for his measuring devices were far too primitive to reveal the tiny angular shifts involved (which are less than an arc-second for even the very closest stars). The first successful parallax determinations came in the years 1837–1839, some 15 years after his death.

The primary target employed by Herschel in his effort to measure stellar parallax was the beautifully brilliant, bluish-white double star Castor (α Geminorum). After 25 years of studying this and other pairs, he announced in another paper to the Royal Society in 1803 that – while he failed to detect any sign of an annual parallactic shift – he had found positive evidence that binary stars are real. In the case of Castor, the fainter companion was slowly but definitely orbiting the brighter primary, proving that the two suns were physically (gravitationally) related. As the first demonstration of the validity of Newton's law of gravity beyond the Solar System, this was an incredibly momentous discovery that electrified the entire scientific world!

Making the Photon Connection

In conclusion, we return to the aesthetic aspect of Herschel's legacy already touched upon above. Sir William was strictly a visual observer of the cosmic pageantry, photography of the heavens lying in the future (his son Sir John being one of its pioneers). This meant that all of his amazing discoveries resulted from actually *looking* at the objects of his study with his eyes. And like him, we stargazers today have an opportunity to not only see and experience the universe firsthand at its grandest with our own eyes, but also to come into actual physical contact with it through the amazing "photon connection."

This is a term coined by the author in the June, 1994, issue of *Sky & Telescope* magazine. It involves the fact that the light by which we see celestial objects like stars, nebulae, and galaxies consists of photons, which have a strange and mysterious dualistic nature. They behave as if they are *both* particles and waves – or particles moving in waves, if you like. Something that was once inside of that object has traveled across the vastness of space and time, and ended its long journey on the retina of your eye. In other words, *you are in direct physical contact with the object you are viewing!* Little wonder that the poet Sarah Teasdale said in looking up at the stars, "I know that I am honored to be witness of so much majesty."

There is no question that William Herschel's profound discoveries and insights in the field of astronomy resulted from a combination of his rare genius, perseverance, abiding love of the heavens, and boundless energy and enthusiasm. But the author is also absolutely convinced that there was yet another factor at play. And it was that as a strictly visual observer, he made the photon connection at the eyepieces of his telescopes night after night over his long career.

Now, reluctantly, we take leave of one of the great shining lights in the long annals of amateur and professional astronomy. But as we do, let us remember that every clear night brings with it an opportunity to retrace the glorious pathway in the sky left by the man known as Sir William Herschel. Shall we not take our telescopes and humbly follow his footsteps?

Appendix 1

Herschel Clubs

As already mentioned in the Preface to this book, the author suggested in a letter in the April, 1976, issue of *Sky & Telescope* magazine (followed by full-length articles both in it and *Astronomy* magazine in which the suggestion was repeated) a way to make viewing William Herschel's catalog of deep-sky wonders more attractive to observers. His 2,508 discoveries were arranged into eight classes, designated I to VIII (see Chapter 3). Of these, 1,893 lie in Classes II and III – his faint and very faint nebulae. Dropping all of those objects as largely more difficult and visually less interesting sights leaves 615 entries in the remaining six classes – which is a much more manageable and realistic (as well as pleasurable!) target list to tackle.

As a result of these published pieces, observation of the Herschel objects began to grow in popularity among the stargazing community, and acting upon my suggestion an actual Herschel Club was started by the Ancient City Astronomy Club in St. Augustine, Florida in 1976. This local effort was soon adopted on a national level in 1980 by the Astronomical League, a federation of most of the astronomy clubs and many independent amateurs in the United States. (See its manual *Observe: The Herschel Objects* by Lydel Guzman, Brenda Guzman, and Paul Jones, first published in 1980, and also the web site below.) Unfortunately, the target list adopted by these two organizations contains only 400 entries rather than the full 615 that the author recommended. Among the 400 objects selected are many in Classes II and III (which in most cases are anything but exciting at the eyepiece), while a number of Herschel showpieces are unfortunately overlooked. A very complete and valuable web site is maintained by the Astronomical League's Herschel Club. It not only offers much helpful information about the observing list itself and requirements for being awarded a coveted Herschel Club Certificate, but also reading material about William Herschel and other related matters. It can be found at: http://astroleague.org/al/obsclubs/herschel/hers400.html.

Since the formation of the original Club, the Rose City Astronomers in Portland, Oregon, has initiated a second Herschel Club list with an additional 400 targets – which the Astronomical League adopted as a Herschel II Club. (See its manual *Observe: The Herschel II Objects*, by Carol Cole and Candace Pratt, published in 1997.) In both cases, the League issues an official certificate of recognition for those observing all 400 objects. Those readers interested in participating in these Herschel observing groups should contact Brenda Branchett, Ancient City Astronomy Club, 515 Glen Haven Drive, Deltona, FL 32738, and Candace Pratt, Rose City Astronomers, Oregon Museum of Science and Industry, 1945 SE Water Street, Portland, OR 97214. And if you really want to see how "Herschelmania" has spread throughout the amateur astronomy community, just enter the words "Herschel Clubs" in your computer's search window. This will bring up a multitude of local astronomy groups across the country (perhaps even one in your area) and abroad that sponsor their own clubs. But should there be no local Herschel Club to become active in, you (or the astronomy group you belong to) might consider forming one of your own – in addition to participating in the two national ones.

As pleased as the author is with the above activities, the 1976 suggestion of forming a Herschel Club may not have been the first. As reported in the February, 1958, issue of

Sky & Telescope, two members of the Royal Astronomical Society of Canada – Dr. T.F. Morris and Tom Noseworthy – had set about observing as many of Herschel's discoveries as possible that long ago. In addition, another Canadian amateur, Patrick Brennan, reported in the August, 1976, issue of the same magazine that he had personally viewed 1,097 Herschel objects over a period of four years. Although these observers apparently never actually formed a Herschel Club as such, they were certainly looking beyond the Messier Clubs of their time.

Mention must also be made here of the William Herschel Society, headquartered at 19, New King Street in Bath, England. This is the only surviving residence of Sir William and his sister Caroline – and it is also where he discovered the planet Uranus. In addition to astronomical, historical, and musical exhibits, visitors can visit the garden from which he observed. Long a private home, the Society purchased this house for preservation as a historic landmark and museum in the late 1970s. The Society's president is the well-known British observer and astronomy popularizer, Sir Patrick Moore. Should you ever be traveling in the UK, you definitely owe it to yourself to pay a visit to this shrine, as well as the many other historical astronomical sites throughout the country. Among these is the Science Museum of London, which has a number of Herschel artifacts and telescopes – including the 6.2-inch aperture 7-foot reflector with which he conducted his first review of the heavens and used to find Uranus.

It was announced in March of 1958 that the contents of Herschel's last residence (known as "Observatory House") in Slough, Buckinghamshire – where he erected his famed 40-foot telescope – were being sold at auction by Sotheby's. This was not only to include the mirrors, eyepieces and parts of Sir William's 20-foot and 40-foot telescopes, and Caroline's instruments as well, but also his books, papers, letters, and musical manuscripts. (Sir John Herschel's pioneering papers on photography were also being put up for auction.) The William Herschel Society continues its efforts to bring as many of these cherished relics as possible back to its Herschel House and Museum in Bath. More information about the Society and its journal *The Speculum* can be found at: http://www.williamherschel.org.uk. Sadly, the hallowed ground on which Observatory House itself once sat is now occupied by a huge office building.

Selected Herschel References

Literally hundreds of books, scholarly papers and articles have been published about the life and work of Sir William, Sir John and Caroline Herschel over the past two centuries, beginning with William's papers in the *Philosophical Transactions* of the Royal Society of London starting in 1786. While on the staff of the Allegheny Observatory in Pittsburgh, Pennsylvania (home of the historic 13-inch Fitz-Clark refractor and the mammoth 30-inch Brasher refractor – fifth largest in the world), the author had the opportunity of examining these original published papers. What a privilege and thrill to read of his discoveries, just as the world first learned of them! Many of the books themselves are long out-of-print and available only in astronomical research libraries (or through used/rare book stores and private collectors), while the papers and articles require accessing the journals and magazines in which they appeared. Below are some of the best Herschel references, a number of which are still currently in print. In several cases the date given for books is that of the most recent edition or printing rather than the original date of publication.

Books and Manuals

Sir William Herschel, E.S. Holden, Charles Scribner's Sons, New York, 1881. An excellent in-depth account of Sir William's life and work – one of the very finest ever written.

The Herschels and Modern Astronomy, Agnes Clerke, Cassell and Company, London, 1895. A charming overview of all three Herschel's and their impact on astronomy of the time.

The Great Astronomers, Sir Robert Ball, Isbister & Company, Ltd., London, 1895. Wonderful account of William Herschel with fascinating illustrations.

Herschel at the Cape, David Evans, University of Texas at Austin, Austin, TX, 1969. A work devoted exclusively to the many discoveries of Sir John Herschel during his famous survey of the southern sky at Cape Town, South Africa.

The History of the Telescope, Henry King, Dover Publications, New York, 2003. Contains one of the best concise accounts of William Herschel's life, telescopes and work in print.

William Herschel and His Work, James Sime, Charles Scribner's Sons, New York, 1900. A work which the author has not seen but which is often referenced.

The Herschel Chronicle: The Life Story of William Herschel and His Sister Caroline Herschel, Constance Lubbock, Cambridge University Press, Cambridge, 1933; reprinted by the William Herschel Society, London, in 1997. One of the most in-depth accounts of the Herschel's ever published.

Scientific Papers of Sir William Herschel, J.L.E. Dreyer, Editor, Royal Society, London, 1912. This compendium, reprinted jointly by the Royal Society and the Royal Astronomical Society, is a most valuable reference for those desiring to read Herschel's original published papers without having access to the *Philosophical Transactions*.

Memoir and Correspondence of Caroline Herschel, Mrs. John Herschel, John Murray, London, 1876. Intimate account of Caroline's life from her many personal letters, as compiled by Sir John Herschel's wife.

The Herschel Partnership: As Viewed by Caroline, Michael Hoskin, Science History Publications, Ltd., London, 2003. A relatively recent book offering fascinating new insights into the working relationship between Sir William and his devoted sister by a noted historian of astronomy.

Caroline Herschel's Autobiographies, Michael Hoskin, Science History Publications, Ltd., London, 2003. Another scholarly work by Hoskin for those wanting to know more about Caroline Herschel herself, as told in her own words.

William Herschel and the Construction of the Heavens, Michael Hoskin, Norton and Norton, New York, 1964. Yet another, earlier scholarly contribution by Hoskin, focusing on Sir William's cosmology.

William Herschel, Explorer of the Heavens, J.B. Sidgwick, Faber and Faber, London, 1953. Comprehensive work by a noted British authority on observational astronomy.

The King's Astronomer, William Herschel, Deborah Crawford, J. Messner, New York, 1968. A work which the author has not seen but which comes well recommended.

New General Catalogue of Nebulae and Clusters of Stars, J.L.E. Dreyer, Royal Astronomical Society, London, 1971. The ultimate reference on all of the discoveries of both William and John Herschel (including those objects actually found by Caroline Herschel).

The Revised New General Catalogue of Nonstellar Astronomical Objects, Jack Sulentic and William Tifft, University of Arizona Press, Tucson, AZ, 1973. The comprehensive update of the original *NGC* which contains the notorious "nonexistent" Herschel objects.

The Bedford Catalogue, A Cycle of Celestial Objects, Volume 2, W.H Smyth, Willmann-Bell, Inc., Richmond Virginia, 1986. This famous early observing guide originally appeared in 1844. It contains detailed and often very quaint descriptions of the visual appearance of many of the brighter Herschel objects using Sir William's designations.

Celestial Objects for Common Telescopes, Volume 2, T.W. Webb, Dover Publications, New York, 1962. The clusters and nebulae described in this classic observing guide originally published in 1859 are given with their original Herschel as well as *NGC* designations. (Also, many of the double and multiple stars listed in it carry William's and John's designations for these objects as well.)

1001 Celestial Wonders, C.E. Barns, Pacific Science Press, Morgan Hill, CA, 1929. This wonderful classic handbook covers every aspect of amateur astronomy, and lists all non-Messier clusters and nebulae by their Herschel classes and numbers.

New Handbook of the Heavens, Hubert Bernhard, Dorothy Bennett and Hugh Rice, McGraw-Hill, New York, 1956. (This work also appeared in many paperback printings under the Mentor/The New American Library imprint up through 1959.) Another superb classic which designates all non-Messier clusters and nebulae in its listing of deep-sky objects by their Herschel classes and numbers where available.

Observe: The Herschel Objects, Brenda and Lydel Guzman and Paul Jones, published by The Astronomical League, East Peoria, IL, 1980. The original manual for observing the 400 Herschel objects selected as targets for the US-based Astronomical League's Herschel Club (see Appendix 1).

Observe: The Herschel II Objects, Carol Cole and Candace Pratt, published by The Astronomical League, West Burlington, IA, 1997. A follow-up manual to the original roster, giving 400 additional Herschel objects selected as targets for the US-based Astronomical League's Herschel Club (see Appendix 1).

Steve O'Meara's Herschel 400 Observing Guide: How to Find & Explore 400 Star Clusters, Nebulae and Galaxies Discovered by Sir William Herschel, Stephen James O'Meara, Cambridge University Press, Cambridge, 2007. Comprehensive coverage of the original Herschel Club's target list of 400 objects written by one of the foremost visual observers in the world today.

Celestial Harvest: 300-Plus Showpieces of the Heavens for Telescope Viewing & Contemplation, James Mullaney, Dover Publications, New York, 2002. This work by the author contains extensive descriptions of the best of the Herschel objects.

Journal and Magazine Articles

"The Telescopes of William Herschel," J.A. Bennett, *Journal for the History of Astronomy*, Volume 7, Number 75, 1976. Scholarly paper for those interested in learning more about the instruments Sir William made and used in his exploration of the heavens.

"William Herschel: Pioneer of the Stars", Brian Jones, *Astronomy*, November, 1988. Great overview of the Herschel story with lots of wonderful illustrations.

"Astronomy's Matriarch", Michael Hoskin, *Sky & Telescope*, May, 2005. This article provides a nice summary of Hoskin's book *The Herschel Partnership* (see above).

"He Broke Through the Barriers of the Skies", N.A. Mackenzie, *Sky & Telescope*, March, 1949. An excellent early article on William Herschel and his work.

"The Discovery of Uranus", J.A. Bennett, *Sky & Telescope*, March, 1981. A superb account of Herschel's life and work, with emphasis on the first planet ever to be discovered.

"Herschel's 'Large 20-foot' Telescope", Joseph Ashbrook, *Sky & Telescope*, September, 1977. Extremely informative article by a noted astronomical historian and late *Sky & Telescope* editor about William Herschel's favorite telescope – the one he used to discover most of his clusters and nebulae.

"A Hole in the Sky", Joseph Ashbrook, *Sky & Telescope*, July, 1974. Fascinating account of William Herschel's unwitting discovery of dark nebulae.

"Portrait of a 40-foot Giant", Brian Warner, *Sky & Telescope*, March, 1986. Brief but good account of Herschel's largest and most famous telescope – his 48-inch reflector.

"William Herschel and His Music", Colin Ronan, *Sky & Telescope*, March, 1981. Some fascinating insights for readers interested in this aspect of Herschel's early life.

"Exploring the Herschel Catalogue," James Mullaney, *Sky & Telescope*, September, 1992. One of several articles written by the author over the years aimed at promoting observation of the Herschel objects among amateur astronomers.

"Observing Herschel Objects," James Mullaney, *Astronomy*, January, 1978. Another, much earlier article by the author on viewing the Herschel objects.

Target List of 615 Herschel Objects

As per the author's 1976 suggestion in *Sky & Telescope* magazine, presented here in the order of increasing Right Ascension (from west to east across the sky) are the 615* objects in William Herschel's Classes I, IV, V, VI, VII, and VIII. (Again, those in Classes II and III are omitted as being largely more difficult and visually less interesting objects). Arranged by Class are the Herschel designation, the corresponding *NGC* number, the constellation (using the standard International Astronomical Union's three-letter abbreviation), the Right Ascension and Declination for Epoch 2000.0, the object's actual type** (which, as discussed in Chapter 3, in many cases does not agree with the Class Herschel assigned it to!), its visual magnitude, angular size in minutes (') or seconds (") of arc, and Messier (M) or Caldwell (C) number plus popular name if any. Bolded entries are those showpieces described in Chapters 5 through 10 (which include a translation of Herschel's shorthand description taken from the *NGC* itself). All of these targets are visible in an 8-inch aperture under good sky conditions and most of them in scopes half that size.

Class I: Bright Nebulae (288)							
H DSG	NGC #	CON	RA & DEC	TYPE	MAG	SIZE	NAME
I-159	278	CAS	00 52 + 47 33	GX	10.9	2′	
I-54	393	CAS	01 09 + 39 40	GX	13p***	–	
I-108	467	PSC	01 19 + 03 18	GX	11.9	2′	
I-151	524	PSC	01 25 + 09 32	GX	10.6	3′	
I-100	584	CET	01 31–06 52	GX	10.4	4′×2′	
I-281	613	SCL	01 34–29 25	GX	10.0	6′×5′	
I-193	**650/1**	**PER**	**01 42 + 51 34**	**PN**	**11.5**	**140″ × 70″**	**=M76/ Little Dumb-bell Nebula**
I-157	672	TRI	01 48 + 27 26	GX	10.8	7′×3′	

*Actually 618 due to dual entries.
**OC = open cluster, GC = globular cluster, PN = planetary nebula, DN = diffuse nebula, SR = supernova remnant, GX = galaxy, NE = "nonexistent."
***The letter "p" indicates a photographic magnitude (rounded to the nearest whole number) rather than a visual one, which averages roughly a magnitude fainter than what the eye sees. The primary data sources for the above table were *Sky Catalogue 2000.0* (Volume 2) and *NGC 2000.0*, while the Herschel designations were taken from the original *NGC*.

Class I: Bright Nebulae (288)—(Continued)

I-62	701	CET	01 51 – 09 42	GX	12.2	2'×1'	
I-105	720	CET	01 53 – 13 44	GX	10.2	4'×3'	
I-112	**772**	**ARI**	**01 59 + 19 01**	**GX**	**10.3**	**7'×4'**	
I-101	779	CET	02 00 – 05 58	GX	11.0	4'×1'	
I-152	821	ARI	02 08 + 11 00	GX	10.8	4'×2'	
I-153	908	CET	02 23–21 14	GX	10.2	6'×3'	
I-154	949	TRI	02 31 + 37 08	GX	11.9	3'×2'	
I-102	1022	CET	02 38 – 06 40	GX	11.4	2'	
I-156	**1023**	**PER**	**02 40 + 39 04**	**GX**	**9.5**	**9'×3'**	
I-63	1052	CET	02 41 – 08 15	GX	10.6	3'×2'	
I-1 = II-6	1055	CET	02 42 + 00 26	GX	10.6	8'×3'	
I-64	**1084**	**ERI**	**02 46 – 07 35**	**GX**	**10.6**	**3'×2'**	
I-109	1201	FOR	03 04 – 26 04	GX	10.6	4'×3'	
I-106	1309	ERI	03 22 – 15 24	GX	11.6	2'	
I-60	1332	ERI	03 26 – 21 20	GX	10.3	5'×2'	
I-257	1344	FOR	03 28 – 31 04	GX	10.3	4'×2'	
I-58	1395	ERI	03 38 – 23 02	GX	11.3	3'×2'	
I-107	**1407**	**ERI**	**03 40 – 18 35**	**GX**	**9.8**	**2'**	
I-155	1453	ERI	03 46 – 03 58	GX	11.6	2'	
I-258	1491	PER	04 03 + 51 19	DN	–	3'	
I-217	1579	PER	04 30 + 35 16	DN	–	12'×8'	
I-158	1600	ERI	04 32 – 05 05	GX	11.1	2'	
I-122	1637	ERI	04 42 – 02 51	GX	10.9	3'	
I-261	1931	AUR	05 31 + 34 15	OC + DN	11.3	3'	
I-218	**2419**	**LYN**	**07 38 + 38 53**	**GC**	**10.4**	**4'**	**=C25/ Inter- galactic Wand- erer**
I-204	2639	UMA	08 44 + 50 12	GX	11.8	2'×1'	
I-200	**2683**	**LYN**	**08 53 + 33 25**	**GX**	**9.7**	**9'×2'**	
I-242	2681	UMA	08 54 + 51 19	GX	10.3	4'	
I-288	2655	CAM	08 56 + 78 13	GX	10.1	5'×4'	
I-249	2742	UMA	09 08 + 60 29	GX	11.7	3'×2'	
I-2	**2775**	**CNC**	**09 10 + 07 02**	**GX**	**10.3**	**4'**	**=C48**
I-250	2768	UMA	09 12 + 60 02	GX	10.0	6'×3'	
I-66	2781	HYA	09 12 – 14 49	GX	11.5	4'×2'	
I-59	2784	HYA	09 12 – 24 10	GX	10.1	5'×2'	
I-167	2782	LYN	09 14 + 40 07	GX	11.5	4'×3'	
I-216	2787	UMA	09 19 + 69 12	GX	10.8	3'×2'	
I-113	2830	LYN	09 20 + 33 44	GX	15p	1'	
I-205	**2841**	**UMA**	**09 22 + 50 58**	**GX**	**9.3**	**8'×4'**	
I-132	2855	HYA	09 22 – 11 55	GX	11.6	3'×2'	
I-137	2859	LMI	09 24 + 34 31	GX	10.7	5'×4'	
I-260	2880	UMA	09 30 + 62 30	GX	11.6	3'×2'	
I-56/57	**2903/5**	**LEO**	**09 32 + 21 30**	**GX**	**8.9**	**13'×7'**	
I-114	2964	LEO	09 43 + 31 51	GX	11.3	3'×2'	
I-61	2974	SEX	09 43 – 03 42	GX	10.8	3'×2'	
I-282	2977	DRA	09 44 + 74 52	GX	13p	2'×1'	
I-285	2976	UMA	09 47 + 67 55	GX	10.2	5'×2'	
I-78	2985	UMA	09 50 + 72 17	GX	10.5	4'×3'	
I-115	3021	LMI	09 51 + 33 33	GX	13p	2'×1'	

(Continued)

Class I: Bright Nebulae (288)—(Continued)

I-286	3077	UMA	10 03 + 68 44	GX	9.8	5'×4'	
I-163	**3115**	**SEX**	**10 05 – 07 43**	**GX**	**9.2**	**8'×3'**	**=C53/ Spindle Galaxy**
I-3	**3166**	**SEX**	**10 14 + 03 26**	**GX**	**10.6**	**5'×3'**	
I-4	**3169**	**SEX**	**10 14 + 03 28**	**GX**	**10.5**	**5'×3'**	
I-79	3147	DRA	10 17 + 73 24	GX	10.6	4'	
I-168	**3184**	**UMA**	**10 18 + 41 25**	**GX**	**9.8**	**7'**	
I-265	3182	UMA	10 20 + 58 12	GX	13p	2'	
I-199	3198	UMA	10 20 + 45 33	GX	10.4	8'×4'	
I-266	3206	UMA	10 22 + 56 56	GX	12p	3'×2'	
I-283	3218	DRA	10 26 + 74 39	NE	–	–	
I-86	3245	LMI	10 27 + 28 30	GX	10.8	3'×2'	
I-72	3254	LMI	10 29 + 29 30	GX	11.5	5'×2'	
I-164	3294	LMI	10 36 + 37 20	GX	11.7	3'×2'	
I-272	3332	LEO	10 40 + 09 11	GX	13p	–	
I-81	3344	LMI	10 44 + 24 55	GX	10.0	7'	
I-26	3345	LEO	10 44 + 11 59	DBL STAR	– –		
I-80	3348	UMA	10 47 + 72 50	GX	11.2	2'	
I-17	**3379**	**LEO**	**10 48 + 12 35**	**GX**	**9.3**	**4'**	**=M105**
I-18	**3384**	**LEO**	**10 48 + 12 38**	**GX**	**10.0**	**6'×3'**	
I-116	3395	LMI	10 50 + 32 59	GX	12.1	2'×1'	
I-117	3396	LMI	10 50 + 32 59	GX	12.2	3'×1'	
I-284	3397	DRA	10 54 + 77 17	NE	–	–	
I-27	3412	LEO	10 51 + 13 25	GX	10.6	4'×2'	
I-118	3430	LMI	10 52 + 32 57	GX	11.5	4'×2'	
I-172	3432	LMI	10 52 + 36 37	GX	11.2	6'×2'	
I-267	3445	UMA	10 55 + 56 59	GX	12.4	2'	
I-233	3448	UMA	10 55 + 54 19	GX	11.7	5'×2'	
I-268	3458	UMA	10 56 + 57 07	GX	13p	2'×1'	
I-87	3486	LMI	11 00 + 28 58	GX	10.3	7'×5'	
I-269	3488	UMA	11 01 + 57 41	GX	14p	2'×1'	
I-88	3504	LMI	11 03 + 27 58	GX	11.1	3'×2'	
I-13	**3521**	**LEO**	**11 06 – 00 02**	**GX**	**8.7**	**10'×5'**	
I-220	3549	UMA	11 11 + 53 23	GX	12p	3'×1'	
I-29	3593	LEO	11 15 + 12 49	GX	11.0	6'×2'	
I-270	3610	UMA	11 18 + 58 47	GX	10.8	3'×2'	
I-241	3621	HYA	11 18 – 32 49	GX	10.6	10'×6'	
I-271	3613	UMA	11 19 + 58 00	GX	12p	4'×2'	
I-244	3619	UMA	11 19 + 57 46	GX	13p	3'	
I-226	3631	UMA	11 21 + 53 10	GX	10.4	5'×4'	
I-245	3642	UMA	11 22 + 59 05	GX	11.1	6'×5'	
I-5	3655	LEO	11 23 + 16 35	GX	11.6	2'×1	
I-20	3666	LEO	11 24 + 11 21	GX	12p	4'×1'	
I-219	3665	UMA	11 25 + 38 46	GX	10.8	3'	
I-131	3672	CRT	11 25 – 09 48	GX	11p	4'×2'	
I-194	3675	UMA	11 26 + 43 35	GX	11p	6'×3'	
I-262	3682	DRA	11 28 + 66 35	GX	13p	2'×1'	
I-246	3683	UMA	11 28 + 56 53	GX	13p	2'×1'	
I-247	3690	UMA	11 28 + 58 33	GX	12p	2'	
I-221	3718	UMA	11 33 + 53 04	GX	10.5	9'×4'	
I-222	3729	UMA	11 34 + 53 08	GX	11.4	3'×2'	

I-287	3735	DRA	11 36 + 70 32	GX	12p	3′×2′	
I-227	3780	UMA	11 39 + 56 16	GX	12p	3′	
I-21	3810	LEO	11 41 + 11 28	GX	10.8	4′×3′	
I-94	3813	UMA	11 41 + 36 33	GX	11.7	2′×1′	
I-201	**3877**	**UMA**	**11 46 + 47 30**	**GX**	**10.9**	**5′×1′**	
I-120	3887	CRT	11 47 − 16 51	GX	11.0	3′	
I-248	3894	UMA	11 49 + 59 25	GX	13p	2′	
I-228	3898	UMA	11 49 + 56 05	GX	10.8	4′×3′	
I-82	3900	LEO	11 49 + 27 01	GX	11.4	4′×2′	
I-259	3923	HYA	11 51 − 28 48	GX	10.1	3′×2′	
I-203	**3938**	**UMA**	**11 53 + 44 07**	**GX**	**10.4**	**5′**	
I-173	3941	UMA	11 53 + 36 59	GX	11p	4′×2′	
I-251	3945	UMA	11 53 + 60 41	GX	10.6	6′×4′	
I-202	3949	UMA	11 54 + 47 52	GX	11.0	3′×2′	
I-67	3962	CRT	11 55 − 13 58	GX	10.6	3′	
I-229	3998	UMA	11 58 + 55 27	GX	10.6	3′×2′	
I-223	4026	UMA	11 59 + 50 58	GX	12p	5′×1′	
I-253	**4036**	**UMA**	**12 01 + 61 54**	**GX**	**10.6**	**4′×2′**	
I-252	**4041**	**UMA**	**12 02 + 62 08**	**GX**	**11.1**	**3′**	
I-174	4062	UMA	12 04 + 31 54	GX	11.2	4′×2′	
I-224	**4085**	**UMA**	**12 05 + 50 21**	**GX**	**12.3**	**3′×1′**	
I-206	**4088**	**UMA**	**12 06 + 50 33**	**GX**	**10.5**	**6′×2′**	
I-207	4096	UMA	12 06 + 47 29	GX	10.6	6′×2′	
I-225	4102	UMA	12 06 + 52 43	GX	12p	3′×2′	
I-195	**4111**	**CVN**	**12 07 + 43 04**	**GX**	**10.8**	**5′×1′**	
I-33 = II-60	4124	VIR	12 08 + 10 23	GX	12p	5′×2′	
I-279	4127	CAM	12 08 + 76 48	GX	12p	2′×1′	
I-263	4128	DRA	12 08 + 68 46	GX	13p	3′×1′	
I-278	4133	DRA	12 09 + 74 56	GX	13p	−	
I-196	4138	CVN	12 10 + 43 41	GX	12p	3′×2′	
I-169	4145	CVN	12 10 + 39 53	GX	11.0	6′×4′	
I-19	**4147**	**COM**	**12 10 + 18 33**	**GC**	**10.3**	**4′**	
I-73	4150	COM	12 11 + 30 24	GX	11.7	2′	
I-165	4151	CVN	12 10 + 39 24	GX	10.4	6′×4′	
I-11	4153	COM	12 11 + 18 22	NE	−	−	
I-208	4157	UMA	12 11 + 50 29	GX	12p	7′×2′	
I-9	**4179**	**VIR**	**12 13 + 01 18**	**GX**	**10.9**	**4′×1′**	
I-175	4203	COM	12 15 + 33 12	GX	10.7	4′×3′	
I-95	4214	CVN	12 16 + 36 20	GX	9.7	8′×6′	
I-35	**4216**	**VIR**	**12 16 + 13 09**	**GX**	**10.0**	**8′×2′**	
I-209	4220	CVN	12 16 + 47 53	GX	12p	4′×2′	
I-74	4245	COM	12 18 + 29 36	GX	11.4	3′	
I-264	4250	DRA	12 17 + 70 48	GX	13p	3′×2′	
I-89	4251	COM	12 18 + 28 10	GX	12p	4′×2′	
I-75	**4274**	**COM**	**12 20 + 29 37**	**GX**	**10.4**	**7′×3′**	
I-90	4278	COM	12 20 + 29 17	GX	10.2	4′	
I-275	4291	DRA	12 20 + 75 22	GX	12p	2′	
I-276	4319	DRA	12 22 + 75 19	GX	12p	3′×2′	
I-139	**4303**	**VIR**	**12 22 + 04 28**	**GX**	**9.7**	**6′**	=M61
I-76	4314	COM	12 23 + 29 53	GX	10.5	5′×4′	
I-277	4386	DRA	12 24 + 75 32	GX	12p	3′×2′	
I-210	4346	CVN	12 24 + 47 00	GX	12p	4′×1′	

(Continued)

Class I: Bright Nebulae (288)—(Continued)

I-65	**4361**	**CRV**	**12 24 – 18 48**	**PN**	**10.3**	**80″**	**=C32**
I-30	4365	VIR	12 24 + 07 19	GX	11p	6′×5′	
I-166	4369	CVN	12 25 + 39 23	GX	12p	2′	
I-22	4371	VIR	12 25 + 11 42	GX	10.8	4′×2′	
I-12	4377	COM	12 25 + 14 46	GX	11.8	2′	
I-123	4378	VIR	12 25 + 04 55	GX	12p	3′	
I-77	4414	COM	12 26 + 31 13	GX	10.3	4′×2′	
I-28.1	**4435**	**VIR**	**12 28 + 13 05**	**GX**	**10.9**	**3′×2′**	The "Eyes" (with I-28.2)
I-28.2	**4438**	**VIR**	**12 28 + 13 01**	**GX**	**10.1**	**9′×4′**	The "Eyes"
I-91	4448	COM	12 28 + 28 37	GX	11.1	4′×2′	
I-213	**4449**	**CVN**	**12 28 + 44 06**	**GX**	**9.4**	**5′×4′**	**=C21**
I-23	4452	VIR	12 29 + 11 45	GX	12.4	2′×1′	
I-161	4459	COM	12 29 + 13 59	GX	10.4	4′×3′	
I-212	4460	CVN	12 29 + 44 52	GX	12p	4′×1′	
I-197	4485	CVN	12 30 + 41 42	GX	12.0	2′	
I-198	**4490**	**CVN**	**12 31 + 41 38**	**GX**	**9.8**	**6′×3′**	Cocoon Galaxy
I-83	4494	COM	12 31 + 25 47	GX	9.9	5′×4′	
I-234	4500	UMA	12 31 + 57 58	GX	13p	2′×1′	
I-31 = I-38	**4526**	**VIR**	**12 34 + 07 42**	**GX**	**9.6**	**7′**	
I-160	4546	VIR	12 36 – 03 48	GX	10.3	4′×2′	
I-36	4550	VIR	12 36 + 12 13	GX	11.6	4′×1′	
I-37	4551	VIR	12 36 + 12 16	GX	11.9	2′	
I-92	**4559**	**COM**	**12 36 + 27 58**	**GX**	**10.0**	**11′×5′**	**=C36**
I-119	4560	VIR	12 36 + 07 41	NE	–	–	
I-273	4589	DRA	12 37 + 74 12	GX	12p	3′	
I-32	4570	VIR	12 37 + 07 15	GX	10.9	4′×1′	
I-124	4580	VIR	12 38 + 05 22	GX	13p	2′	
I-125	4586	VIR	12 38 + 04 19	GX	11.6	4′×2′	
I-43	**4594**	**VIR**	**12 40 – 11 37**	**GX**	**8.3**	**9′×4′**	=M104/ Sombrero Galaxy
I-24	**4596**	**VIR**	**12 40 + 10 11**	**GX**	**10.5**	**4′×3′**	
I-254	4605	UMA	12 40 + 61 37	GX	11.0	6′×2′	
I-178 = I-179	4618	CVN	12 42 + 41 09	GX	10.8	4′	
I-14	4632	VIR	12 42 – 05 05	GX	12p	3′×1′	
I-274	4648	DRA	12 42 + 74 25	GX	13p	2′	
I-10	4643	VIR	12 43 + 01 59	GX	10.6	3′	
I-176/ 177	**4656/7**	**CVN**	**12 44 + 32 10**	**GX**	**10.4**	**14′×3′**	Hockey Stick Galaxy
I-142	4665	VIR	12 45 + 03 03	GX	12p	4′	
I-15	4666	VIR	12 45 – 00 28	GX	10.8	4′×2′	
I-39	**4697**	**VIR**	**12 49 – 05 48**	**GX**	**9.2**	**7′×5′**	**=C52**
I-8	4698	VIR	12 48 + 08 29	GX	10.7	4′×2′	
I-129	4699	VIR	12 49 – 08 40	GX	9.6	4′×3′	

I-140	4713	VIR	12 50 + 05 19	GX	11.8	3'×2'	
I-84	**4725**	**COM**	**12 50 + 25 30**	**GX**	**9.2**	**11'×8'**	
I-41	4731	VIR	12 51 − 06 24	GX	11p	6'×3'	
I-133	4742	VIR	12 52 − 10 27	GX	11.1	2'	
I-16	4753	VIR	12 52 − 01 12	GX	9.9	5'×3'	
I-25	**4754**	**VIR**	**12 52 + 11 19**	**GX**	**10.6**	**5'×3'**	
I-134	4781	VIR	12 54 − 10 32	GX	12p	4'×2'	
I-135	4782	CRV	12 55 − 12 34	GX	11.7	2'	
I-136	4783	CRV	12 55 − 12 34	GX	11.8	2'	
I-93	4793	COM	12 55 + 28 56	GX	11.7	3'×2'	
I-211	4800	CVN	12 55 + 46 32	GX	12p	2'×1'	
I-243	4814	UMA	12 55 + 58 21	GX	13p	3'×2'	
I-141	4808	VIR	12 56 + 04 18	GX	12p	3'×1'	
I-68	4856	VIR	12 59 − 15 02	GX	10.4	5'×2'	
I-162	4866	VIR	13 00 + 14 10	GX	11.0	6'×2'	
I-143	4900	VIR	13 01 + 02 30	GX	11.5	2'	
I-69	4902	VIR	13 01 − 14 31	GX	11.2	3'	
I-40	4941	VIR	13 04 − 05 33	GX	11.1	4'×2'	
I-130	4958	VIR	13 06 − 08 01	GX	10.5	4'×1'	
I-42	4995	VIR	13 10 − 07 50	GX	11.0	2'	
I-96	**5005**	**CVN**	**13 11 + 37 03**	**GX**	**9.8**	**5'×3'**	=C29
I-85	5012	COM	13 12 + 22 55	GX	13p	3'×2'	
I-97	5033	CVN	13 13 + 36 36	GX	10.1	10'×6'	
I-138	5061	HYA	13 18 − 26 50	GX	12p	3'×2'	
I-186	**5195**	**CVN**	**13 30 + 47 16**	**GX**	**9.6**	**5'×4'**	Companion to M51
I-34	**5248**	**BOO**	**13 38 + 08 53**	**GX**	**10.3**	**6'×5'**	=C45
I-98	5273	CVN	13 42 + 35 39	GX	11.6	3'	
I-170	5290	CVN	13 45 + 41 43	GX	13p	4'×1'	
I-180	5297	CVN	13 46 + 43 52	GX	12p	6'×1'	
I-255	5308	UMA	13 47 + 60 58	GX	11.3	4'×1'	
I-256	5322	UMA	13 49 + 60 12	GX	10.0	6'×4'	
I-238	5376	UMA	13 55 + 59 30	GX	13p	2'×1'	
I-6	5363	VIR	13 56 + 05 15	GX	10.2	4'×3'	
I-187	5377	CVN	13 56 + 47 14	GX	11.2	5'×3'	
I-239	5379	UMA	13 56 + 59 45	GX	14p	2'×1'	
I-240	5389	UMA	13 56 + 59 44	GX	13p	4'×1'	
I-181	5383	CVN	13 57 + 41 51	GX	11.4	4'×3'	
I-191	5394	CVN	13 59 + 37 27	GX	13.0	2'×1'	
I-190	5395	CVN	13 59 + 37 25	GX	11.6	3'×2'	
I-230	5422	UMA	14 01 + 55 10	GX	13p	4'	
I-231	**5473**	**UMA**	**14 05 + 54 54**	**GX**	**11.4**	**3'×2'**	
I-214	5474	UMA	14 05 + 53 40	GX	10.9	4'	
I-232	5485	UMA	14 07 + 55 00	GX	11.5	3'×2'	
I-99	5557	BOO	14 18 + 36 30	GX	11.1	2'	
I-235	5585	UMA	14 20 + 56 44	GX	10.9	6'×4'	
I-144	5566	VIR	14 20 + 03 56	GX	10.5	6'×2'	
I-145	5574	VIR	14 21 + 03 14	GX	12.4	2'×1'	
I-146	5576	VIR	14 21 + 03 16	GX	10.9	3'×2'	
I-236	5631	UMA	14 27 + 56 35	GX	13p	2'	
I-185	5633	BOO	14 28 + 46 09	GX	12.3	2'×1'	
I-70	**5634**	**VIR**	**14 30 − 05 59**	**GC**	**9.6**	**5'**	
I-237	5678	DRA	14 32 + 57 55	GX	12p	3'×2'	

(Continued)

Class I: Bright Nebulae (288)—(Continued)

H DSG	NGC #	CON	RA & DEC	TYPE	MAG	SIZE	NAME
I-189	5676	BOO	14 33 + 49 28	GX	10.9	4'×2'	
I-188	5689	BOO	14 36 + 48 45	GX	11.9	4'×1'	
I-182	5713	VIR	14 40 – 00 17	GX	11.4	3'×2'	
I-184	5728	LIB	14 42 – 17 15	GX	11.3	3'×2'	
I-171	5739	BOO	14 42 + 41 50	GX	13p	2'	
I-126	**5746**	**VIR**	**14 45 + 01 57**	**GX**	**10.6**	**8'×2'**	
I-183	5750	VIR	14 46 – 00 13	GX	11.6	3'×2'	
I-71	5812	LIB	15 01 – 07 27	GX	11.2	2'	
I-127	5813	VIR	15 01 + 01 42	GX	10.7	4'×3'	
I-128	5846	VIR	15 06 + 01 36	GX	10.2	3'	
I-215	**5866**	**DRA**	**15 07 + 55 46**	**GX**	**10.0**	**5'×2'**	**Formerly M102**
I-148	5921	SER	15 22 + 05 04	GX	10.8	5'×4'	
I-280	6217	UMI	16 33 + 78 12	GX	11.2	3'	
I-147	6304	OPH	17 14 – 29 28	GC	8.4	7'	
I-45	6316	OPH	17 17 – 28 08	GC	9.0	5'	
I-149	6342	OPH	17 21 – 19 35	GC	9.9	3'	
I-46	6355	OPH	17 24 – 26 21	GC	9.6	5'	
I-48	**6356**	**OPH**	**17 24 – 17 49**	**GC**	**8.4**	**7'**	
I-44	6401	OPH	17 39 – 23 55	GC	9.5	6'	
I-150	**6440**	**SGR**	**17 49 – 20 22**	**GC**	**9.7**	**5'**	
I-49	6522	SGR	18 04 – 30 02	GC	8.6	6'	
I-50	6624	SGR	18 24 – 30 22	GC	8.3	6'	
I-51	6638	SGR	18 31 – 25 30	GC	9.2	5'	
I-47	**6712**	**SCT**	**18 53 – 08 42**	**GC**	**8.2**	**7'**	
I-103	**6934**	**DEL**	**20 34 + 07 24**	**GC**	**8.8**	**7'**	**=C47**
I-52	**7006**	**DEL**	**21 02 + 16 11**	**GC**	**10.6**	**4'**	**=C42**
I-192	7008	CYG	21 01 + 54 33	PN	83"	12?	
I-53	**7331**	**PEG**	**22 37 + 34 25**	**GX**	**9.5**	**11'×4'**	**=C30**
I-55	**7479**	**PEG**	**23 05 + 12 19**	**GX**	**11.0**	**4'×3'**	**=C44**
I-104	7606	AQR	23 19 – 08 29	GX	10.8	6'×3'	
I-110	7723	AQR	23 39 – 12 58	GX	11.1	4'×3'	
I-111	7727	AQR	23 40 – 12 18	GX	10.7	4'×3'	

Class IV: Planetary Nebulae (80)

H DSG	NGC #	CON	RA & DEC	TYPE	MAG	SIZE	NAME
IV-15	16	PEG	00 09 + 27 44	GX	12.0	2'×1'	
IV-58	**40**	**CEP**	**00 13 + 72 32**	**PN**	**10.2**	**60"×40"**	**=C2**
IV-42	676	PSC	01 49 + 05 54	GX	11p	4'×2'	
IV-23	936	CET	02 28 – 01 09	GX	10.1	5'×4'	
IV-43	1186	PER	03 06 + 42 50	GX	13p	3'×1'	
IV-17	1253	ERI	03 14 – 02 49	GX	12p	5'×3'	
IV-77	1325	ERI	03 25 – 21 20	GX	12.8	3'×2'	
IV-53	**1501**	**CAM**	**04 07 + 60 55**	**PN**	**11.9**	**55" × 48"**	**Oyster Nebula**
IV-69	**1514**	**TAU**	**04 09 + 30 47**	**PN**	**10.9**	**2'**	
IV-26	**1535**	**ERI**	**04 14 – 12 44**	**PN**	**9.4**	**20" × 17"**	**Lassell's Extraordinary Object**
IV-32	1700	ERI	04 57 – 04 52	GX	11.0	3'×2'	
IV-21	1964	LEP	05 33 – 21 57	GX	10.8	6'×2'	

IV-33	1999	ORI	05 36 – 06 42	DN	–	16'×12'	
IV-34	**2022**	**ORI**	**05 42 + 09 05**	**PN**	**12.0**	**18"**	
IV-24	2023	ORI	05 42 – 02 14	DN	–	10'	
IV-36	2071	ORI	05 47 + 00 18	DN	–	4'×3'	
IV-44	2167	MON	06 07 – 06 12	NE	–	–	
IV-19	2170	MON	06 08 – 06 24	DN	–	2'	
IV-38	2182	MON	06 10 – 06 20	DN	–	3'	
IV-20	2185	MON	06 11 – 06 13	DN	–	3'	
IV-3	2245	MON	06 33 + 10 10	DN	–	5'×3'	
IV-2	**2261**	**MON**	**06 39 + 08 44**	**DN**	**10.0**	**2'×1'**	**=C46/ Hubble's Variable Neb.**
IV-25	2327	CMA	07 04 – 11 18	DN	–	–	
IV-65	2346	MON	07 09 – 00 48	PN	–	55"	
IV-45	**2392**	**GEM**	**07 29 + 20 55**	**PN**	**8.3**	**20"**	**=C39/ Eskimo/ Clown Face Neb**
IV-39	**2438**	**PUP**	**07 42 – 14 44**	**PN**	**11.0**	**66"**	**In M46**
IV-64	**2440**	**PUP**	**07 42 – 18 13**	**PN**	**10.5**	**16"**	
IV-22	2467	PUP	07 52 – 26 24	DN	8'×7'		
IV-55	2537	LYN	08 13 + 46 00	GX	11.7	2'	
IV-35	2610	HYA	08 33 – 16 09	PN	14p	37"	
IV-66	2701	UMA	08 59 + 53 46	GX	12.4	2'	
IV-68	2950	UMA	09 43 + 58 51	GX	11.0	3'×2'	
IV-79	**3034**	**UMA**	**09 56 + 69 41**	**GX**	**8.4**	**11'×5'**	**=M82**
IV-48	3104	LMI	10 04 + 40 45	GX	13p	3'×2'	
IV-10	3239	LEO	10 25 + 17 10	GX	12p	5'×4'	
IV-27	**3242**	**HYA**	**10 25 – 18 38**	**PN**	**8.6**	**40"×35"**	**=C59/ Jupiter's Ghost**
IV-60	3310	UMA	10 39 + 53 30	GX	10.9	4'×3'	
IV-6	3423	SEX	10 51 + 05 50	GX	11.2	4'	
IV-29	3456	CRT	10 54 – 16 02	GX	13p	2'	
IV-7	3507	LEO	11 04 + 18 08	GX	11p	4'×3'	
IV-59	3658	UMA	11 24 + 38 34	GX	13p	2'	
IV-4	3662	LEO	11 24 – 01 06	GX	14p	2'×1'	
IV-67	3963	UMA	11 55 + 58 30	GX	12p	3'	
IV-62	3982	UMA	11 56 + 55 08	GX	12p	2'	
IV-61	**3992**	**UMA**	**11 58 + 53 23**	**GX**	**9.8**	**8'×5'**	**=M109**
IV-28.1	**4038**	**CRV**	**12 02 – 18 52**	**GX**	**10.7**	**3'×2'**	**=C60/ Anten- nae/**
IV-28.2	**4039**	**CRV**	**12 02 – 1852**	**GX**	**13p**	**3'×2'**	**=C61/ Ring-Tail Galaxy**
IV-56	4051	UMA	12 03 + 44 32	GX	10.3	5'×4'	
IV-54	4143	CVN	12 10 + 42 32	GX	12p	3'×2'	
IV-5	4517	VIR	12 33 + 00 07	GX	10.5	10'×2'	
IV-8	**4567**	**VIR**	**12 36 + 11 15**	**GX**	**11.3**	**3'×2'**	**Siamese Twins (with IV-9)**

(Continued)

Class IV: Planetary Nebulae (80)—(Continued)

IV-9	**4568**	**VIR**	**12 37 + 11 14**	**GX**	**10.8**	**5'×2'**	**Siamese Twins (with IV-8)**
IV-78	4750	DRA	12 50 + 72 52	GX	112p	2'	
IV-40	4804	CRV	12 56 – 13 02	NE	–	–	
IV-30	4861	CVN	12 59 + 34 52	GX	12.2	4'×2'	
IV-47	4915	VIR	13 02 – 04 33	GX	11.9	2'×1'	
IV-70	5144	UMI	13 23 + 70 31	GX	113p	1'	
IV-63	5204	UMA	13 30 + 58 25	GX	11.3	5'×3'	
IV-46	5493	VIR	14 12 – 05 03	GX	11.5	2'×1'	
IV-49	5507	VIR	14 13 – 03 09	GX	13p	2'×1'	
IV-71	5856	BOO	15 07 + 18 27	NE	–	–	
IV-50	**6229**	**HER**	**16 47 + 47 32**	**GC**	**9.4**	**4'**	
IV-57	6301	HER	17 09 + 42 20	GX	14p	–	
IV-11	**6369**	**OPH**	**17 29 – 23 46**	**PN**	**11.5**	**30"**	**Little Ghost Nebula**
IV-41	**6514**	**SGR**	**18 03 – 23 02**	**DN**	**6.3**	**28'**	**=M20/ Trifid Nebula**
IV-37	**6543**	**DRA**	**17 59 + 66 38**	**PN**	**8.8**	**22" × 16"**	**=C6/ Cat's Eye/ Snail Nebula**
IV-12	6553	SGR	18 09 – 25 54	GC	8.2	8'	
IV-14	6772	AQL	19 15 – 02 42	PN	14p	62"	
IV-51	**6818**	**SGR**	**19 44 – 14 09**	**PN**	**9.9**	**22" × 15"**	**Little Gem Nebula**
IV-73	**6826**	**CYG**	**19 45 + 50 31**	**PN**	**8.9**	**27"**	**=C15/ Blinking Plane- tary**
IV-72	**6888**	**CYG**	**20 12 + 38 21**	**DN**	**8.8**	**18'×13'**	**=C27/ Crescent Nebula**
IV-13	6894	CYG	20 16 + 30 34	PN	14p	42"	
IV-16	**6905**	**DEL**	**20 22 + 20 07**	**PN**	**11.9**	**44" × 38"**	**Blue Flash Nebula**
IV-76	**6946**	**CEP**	**20 35 + 60 09**	**GX**	**8.9**	**11'×10'**	**=C12**
IV-74	**7023**	**CEP**	**21 02 + 68 12**	**DN**	**6.8**	**18'**	**=C4/Iris Nebula**
IV-1	**7009**	**AQR**	**21 04 – 11 22**	**PN**	**8.3**	**25" × 17"**	**=C55/ Saturn Nebula**
IV-75	7129	CEP	21 44 + 66 10	DN	–	8'×7'	
IV-31	7302	AQR	22 32 – 14 07	GX	12.1	2'×1'	
IV-52	**7635**	**CAS**	**23 21 + 61 12**	**DN**	**7.0**	**15'×8'**	**=C11/ Bubble Nebula**

H DSG	NGC #	CON	RA & DEC	TYPE	MAG	SIZE	NAME
IV-18	7662	AND	23 26 + 42 33	PN	8.5	32" × 28"	=C22/ Blue Snowball

Class V: Very Large Nebulae (52)

H DSG	NGC #	CON	RA & DEC	TYPE	MAG	SIZE	NAME
V-16	68	AND	00 18 + 30 04	GX	13.0	2'×1'	
V-18	205	AND	00 40 + 41 41	GX	8.0	17'×10'	=M110/ Companion to M31
V-36	206	AND	00 41 + 40 44	OC + DN	–	–	
V-25	246	CET	00 47 – 11 53	PN	8.5	4'	=C56
V-20	247	CET	00 47 – 20 46	GX	8.9	20'×7'	=C62
V-1	253	SCL	00 48 – 25 17	GX	7.1	25'×7'	=C65/ Sculptor Galaxy
V-17	598	TRI	01 34 + 30 39	GX	5.7	62'×39'	=M33/ TRI/Pinwheel Galaxy
V-19	891	AND	02 23 + 42 42	GX	9.9	14'×3'	=C23
V-48	1097	FOR	02 46 – 30 17	GX	9.2	9'×7'	=C67
V-49	1624	PER	04 40 + 50 27	DN	–	5'	
V-32	1788	ORI	05 07 – 03 21	DN	–	8'×5'	
V-33	1908	ORI	05 26 – 02 32	NE	–	–	
V-38	1909	ORI	05 26 – 08 08	NE	–	–	
V-31	1980	ORI	05 35 – 05 24	DN	–	14'	
V-30	1977	ORI	05 36 – 04 52	DN	–	40'×20'	
V-34	1990	ORI	05 36 – 01 12	DN	–	50'	
V-28	2024	ORI	05 41 – 02 27	DN	–	30'	Flame/ Burning Bush Nebula
V-27 = VIII-5	2264	MON	06 41 + 09 54	DN	–	60'×30'	
V-21	2359	CMA	07 19 – 13 12	DN	–	8'×6'	
V-44	2403	CAM	07 37 + 65 36	GX	8.4	18'×11'	=C7
V-50	2997	ANT	09 45 – 31 11	GX	10.6	8'×6'	
V-26	3003	LMI	09 49 + 33 25	GX	11.7	6'×2'	
V-23	3027	UMA	09 56 + 72 12	GX	12p	5'×2'	
V-47	3079	UMA	10 02 + 55 41	GX	10.6	8'×2'	
V-7	3346	LEO	10 44 + 14 52	GX	12p	3'×2'	
V-52	3359	UMA	10 47 + 63 13	GX	10.4	7'×4'	
V-39	3511	CRT	11 03 – 23 05	GX	12p	5'×2'	
V-40	3513	CRT	11 04 – 23 15	GX	12p	3'×2'	
V-46	3556	UMA	11 12 + 55 40	GX	10.1	8'×2'	=M108
V-8	3628	LEO	11 20 + 13 36	GX	9.5	15'×4'	
V-45	3953	UMA	11 54 + 52 20	GX	10.0	7'×4'	
V-4	4123	VIR	12 08 + 02 53	GX	11.2	4'	
V-51	4236	DRA	12 17 + 69 28	GX	9.6	20'×8'	=C3
V-41	4244	CVN	12 18 + 37 49	GX	10.2	16'×2'	=C26
V-43	4258	CVN	12 19 + 47 18	GX	8.3	18'×8'	=M106

(Continued)

Class V: Very Large Nebulae (52)—(Continued)

H DSG	NGC #	CON	RA & DEC	TYPE	MAG	SIZE	NAME
V-5	4293	COM	12 21 + 18 23	GX	11p	6'×3'	
V-29.1	4395	CVN	12 26 + 33 33	GX	10.2	13'×11'	
V-29.2	4401	CVN	12 26 + 33 31	NE	–	–	
V-2	4536	VIR	12 34 + 02 11	GX	10.4	7'×4'	
V-24	**4565**	**COM**	**12 36 + 25 29**	**GX**	**9.6**	**16'×3'**	**=C38**
V-42	**4631**	**CVN**	**12 42 + 32 32**	**GX**	**9.3**	**15'×3'**	**=C32 Hump- back Whale Gal.**
V-3	4910	VIR	13 01 + 01 40	NE	–	–	
V-22	5170	VIR	13 30 – 17 58	GX	12p	8'×1'	
V-6	5293	BOO	13 47 + 16 16	GX	14p	2'	
V-10/ 11/12	**6514**	**SGR**	**18 02 – 23 02**	**OC?**	**6.3**	**28'**	**In M20 Trifid Nebula**
V-9	6526	SGR	18 03 – 23 35	DN	–	40'	
V-13	6533	SGR	18 05 – 24 53	NE	–	–	
V-15	**6960**	**CYG**	**20 46 + 30 43**	**SR**	**7.9**	**70'×6'**	**=C34/ Veil/Fila- mentary Nebula**
V-14	**6992/5**	**CYG**	**20 56 + 31 43**	**SR**	**7.5**	**60'×8'**	**=C33/ Veil/Fila- mentary Nebula**
V-37	**7000**	**CYG**	**20 59 + 44 20**	**DN**	**5.0**	**100'× 60'**	**=C20/ North America Nebula**

Class VI: Very Compressed and Rich Clusters of Stars (42)

H DSG	NGC #	CON	RA & DEC	TYPE	MAG	SIZE	NAME
VI-35	136	CAS	00 32 + 61 32	OC	–	1'	
VI-20	**288**	**SCL**	**00 53 – 26 35**	**GC**	**8.1**	**14'**	
VI-31	**663**	**CAS**	**01 46 + 61 15**	**OC**	**7.1**	**16'**	**=C10**
VI-33	**869**	**PER**	**02 19 + 57 09**	**OC**	**3.5**	**30'**	**=C14/ Double Cluster**
VI-34	**884**	**PER**	**02 22 + 57 07**	**OC**	**3.6**	**30'**	**=C14/ Double Cluster**
VI-25	**1245**	**PER**	**03 15 + 47 15**	**OC**	**8.4**	**10'**	
VI-26	1605	PER	04 35 + 45 15	OC	10.7	5'	
VI-17	**2158**	**GEM**	**06 08 + 24 06**	**OC**	**8.6**	**5'**	
VI-5	**2194**	**ORI**	**06 14 + 12 48**	**OC**	**8.5**	**10'**	
VI-28	2259	MON	06 39 + 10 53	OC	11p	4'	
VI-21	**2266**	**GEM**	**06 43 + 26 58**	**OC**	**9.8**	**7'**	
VI-3	2269	MON	06 44 + 04 34	OC	10.0	4'	
VI-27	**2301**	**MON**	**06 52 + 00 28**	**OC**	**6.0**	**12'**	
VI-2	2304	GEM	06 55 + 18 01	OC	10p	5'	
VI-18	2309	MON	06 56 – 07 12	OC	11p	3'	

H DSG	NGC #	CON	RA & DEC	TYPE	MAG	SIZE	NAME
VI-6	2355	GEM	07 17 + 13 47	OC	10p	9'	
VI-1	**2420**	**GEM**	**07 39 + 21 34**	**OC**	**8.3**	**10'**	
VI-36	2432	PUP	07 41 – 19 05	OC	10p	8'	
VI-37	**2506**	**MON**	**08 00 – 10 46**	**OC**	**7.6**	**12'**	**=C54**
VI-22	**2548**	**HYA**	**08 14 – 05 48**	**OC**	**5.8**	**30'**	**=M48**
VI-39	2571	PUP	08 19 – 29 44	OC	7.0	13'	
VI-4	3055	SEX	09 55 + 04 16	GX	12.1	2'×1'	
VI-7	**5053**	**COM**	**13 16 + 17 42**	**GC**	**9.8**	**10'**	
VI-9	**5466**	**BOO**	**14 06 + 28 32**	**GC**	**9.1**	**11'**	
VI-19 =	**5897**	**LIB**	**15 17 – 21 01**	**GC**	**8.6**	**13'**	
VI-8?							
VI-10	**6144**	**SCO**	**16 27 – 26 02**	**GC**	**9.1**	**9'**	
VI-40	**6171**	**OPH**	**16 32 – 13 03**	**GC**	**8.1**	**10'**	**=M107**
VI-11	6284	OPH	17 04 – 24 46	GC	9.0	6'	
VI-12	6293	OPH	17 10 – 26 35	GC	8.2	8'	
VI-41	6412	DRA	17 30 + 75 42	GX	11.8	2'	
VI-13	6451	SCO	17 51 – 30 13	OC	8.2	8'	
VI-23	**6645**	**SGR**	**18 33 – 16 54**	**OC**	**8.5**	**10'**	
VI-15	6678	DRA	18 33 + 67 51	NE	–	–	
VI-14	6802	VUL	19 31 + 20 16	OC	8.8	3'	
VI-38	6804	AQL	19 32 + 09 13	PN	12p	1'	
VI-16	6839	SGE	19 55 + 17 54	NE	–	–	
VI-42	**6939**	**CEP**	**20 31 + 60 38**	**OC**	**7.8**	**8'**	
VI-24	7044	CYG	21 13 + 42 29	OC	–	–	
VI-32	7086	CYG	21 30 + 51 35	OC	8.4	9'	
VI-29	7245	LAC	22 15 + 54 20	OC	9.2	5'	
VI-30	**7789**	**CAS**	**23 57 + 56 44**	**OC**	**6.7**	**16'**	**Caroline's Cluster**

Class VII: Compressed Clusters of Small and Large Stars (67)

H DSG	NGC #	CON	RA & DEC	TYPE	MAG	SIZE	NAME
VII-45	436	CAS	01 16 + 58 49	OC	8.8	6'	
VII-42	**457**	**CAS**	**01 19 + 58 20**	**OC**	**6.4**	**13'**	**=C13/ Owl/ET Cluster**
VII-48	**559**	**CAS**	**01 30 + 63 18**	**OC**	**9.5**	**7'**	**=C8**
VII-49	637	CAS	01 43 + 64 00	OC	8.2	4'	
VII-46	654	CAS	01 44 + 61 53	OC	6.5	5'	
VII-32	**752**	**AND**	**01 58 + 37 50**	**OC**	**5.7**	**50'**	**=C28**
VII-3	1498	ERI	04 00 – 12 01	NE	–	–	
VII-47	**1502**	**CAM**	**04 08 + 62 20**	**OC**	**5.7**	**8'**	**Golden Harp Cluster**
VII-60	1513	PER	04 10 + 49 31	OC	8.4	9'	
VII-61	**1528**	**PER**	**04 15 + 51 14**	**OC**	**6.4**	**24'**	
VII-1	1662	ORI	04 48 + 10 56	OC	6.4	20'	
VII-21	**1758**	**TAU**	**05 04 + 23 49**	**OC**	**7.0**	**42'**	
VII-4	**1817**	**TAU**	**05 12 + 16 42**	**OC**	**7.7**	**16'**	
VII-33	**1857**	**AUR**	**05 20 + 39 21**	**OC**	**7.0**	**6'**	
VII-34	1883	AUR	05 26 + 46 33	OC	12p	2'	
VII-39	1907	AUR	05 28 + 35 19	OC	8.2	7'	
VII-24	2112	ORI	05 54 + 00 24	OC	9p	11'	
VII-25	2186	ORI	06 12 + 05 27	OC	8.7	4'	
VII-57	2192	AUR	06 15 + 39 51	OC	11p	6'	

(Continued)

Class VII: Compressed Clusters of Small and Large Stars (67)—(Continued)							
H DSG	NGC #	CON	RA & DEC	TYPE	MAG	SIZE	NAME
VII-13	2204	CMA	06 16 – 18 39	OC	9.6	13′	
VII-20	2215	MON	06 21 – 07 17	OC	8.4	11′	
VII-35	2224	GEM	06 28 + 12 38	NE	–	–	
VII-26	2225	MON	06 27 – 09 39	OC	–	–	
VII-5	2236	MON	06 30 + 06 50	OC	8.5	7′	
VII-2	**2244**	**MON**	**06 32 + 04 52**	**OC**	**4.8**	**24′**	**=C50/ Rosette Cluster**
VII-54	2253	CAM	06 42 + 66 20	OC	–	–	
VII-22	2254	MON	06 36 + 07 40	OC	9.7	4′	
VII-37	2262	MON	06 38 + 01 11	OC	11p	4′	
VII-36	2270	MON	06 44 + 03 26	NE	–	–	
VII-14	2318	CMA	07 00 – 13 42	OC	–	–	
VII-38	2324	MON	07 04 + 01 03	OC	8.4	8′	
VII-27	2349	MON	07 10 – 08 37	NE	–	–	
VII-15	2352	CMA	07 14 – 24 06	NE	–	–	
VII-16	2354	CMA	07 14 – 25 44	OC	6.5	20′	
VII-6	2356	GEM	07 17 + 13 58	NE	–	–	
VII-12	**2360**	**CMA**	**07 18 – 15 37**	**OC**	**7.2**	**13′**	**=C58**
VII-17	**2362**	**CMA**	**07 19 – 24 57**	**OC**	**4.1**	**8′**	**=C64/ Tau CMA Cluster**
VII-65	2401	PUP	07 29 – 13 58	OC	13p	2′	
VII-67	2421	PUP	07 36 – 20 37	OC	8.3	10′	
VII-28	2423	PUP	07 37 – 13 52	OC	6.7	19′	
VII-58	2479	PUP	07 55 – 17 43	OC	10p	7′	
VII-10	2482	PUP	07 55 – 24 18	OC	7.3	12′	
VII-23	2489	PUP	07 56 – 30 04	OC	7.9	8′	
VII-11	**2539**	**PUP**	**08 11 – 12 50**	**OC**	**6.5**	**22′**	
VII-64	**2567**	**PUP**	**08 19 – 30 38**	**OC**	**7.4**	**10′**	
VII-63	2627	PYX	08 37 – 29 57	OC	8p	11′	
VII-29	5998	SCO	15 49 – 28 36	NE	–	–	
VII-7	**6520**	**SGR**	**18 03 – 27 54**	**OC**	**8.1**	**6′**	
VII-30	6568	SGR	18 13 – 21 36	OC	9p	13′	
VII-31	6583	SGR	18 16 – 22 08	OC	10p	3′	
VII-19	**6755**	**AQL**	**19 08 + 04 14**	**OC**	**7.5**	**15′**	
VII-62	6756	AQL	19 09 + 04 41	OC	11p	4′	
VII-18	6823	VUL	19 43 + 23 18	OC	7.1	12′	
VII-9	6830	VUL	19 51 + 23 04	OC	7.9	12′	
VII-59	**6866**	**CYG**	**20 04 + 44 00**	**OC**	**7.6**	**7′**	**The Kite Cluster**
VII-8	**6940**	**VUL**	**20 35 + 28 18**	**OC**	**6.3**	**31′**	
VII-51	7062	CYG	21 23 + 46 23	OC	8.3	7′	
VII-50	7067	CYG	21 24 + 48 01	OC	9.7	3′	
VII-52	7082	CYG	21 29 + 47 05	OC	7.2	25′	
VII-40	7128	CYG	21 44 + 53 43	OC	9.7	3′	
VII-66	7142	CEP	21 46 + 65 48	OC	9.3	4′	
VII-53	7209	LAC	22 05 + 46 30	OC	6.7	25′	
VII-41	7296	LAC	22 28 + 52 17	OC	10p	4′	
VII-43	7419	CEP	22 54 + 60 50	OC	13p	2′	
VII-44	**7510**	**CEP**	**23 12 + 60 34**	**OC**	**7.9**	**4′**	
VII-55	7762	CEP	23 50 + 68 02	OC	10p	11′	
VII-56	7790	CAS	23 58 + 61 13	OC	8.5	17′	

Class VIII: Coarsely Scattered Clusters of Stars (89)							
VIII-29	7826	CET	00 05 – 20 44	NE	–	–	
VIII-59	129	CAS	00 30 + 60 14	OC	6.5	21′	
VIII-78	**225**	**CAS**	**00 43 + 61 47**	**OC**	**7.0**	**12′**	
VIII-64	381	CAS	01 08 + 61 35	OC	9p	6′	
VIII-66	1027	CAS	02 43 + 61 33	OC	6.7	20′	
VIII-65	659	CAS	01 44 + 60 42	OC	7.9	5′	
VIII-88	1342	PER	03 32 + 37 20	OC	6.7	14′	
VIII-84	1348	PER	03 34 + 51 26	OC	–	–	
VIII-80	1444	PER	03 49 + 52 40	OC	6.6	4′	
VIII-85	1545	PER	04 21 + 50 15	OC	6.2	18′	
VIII-70	1582	PER	04 32 + 43 51	OC	7p	37′	
VIII-8	**1647**	**TAU**	**04 46 + 19 04**	**OC**	**6.4**	**45′**	
VIII-7	1663	ORI	04 49 + 13 10	OC	–	–	
VIII-59	1664	AUR	04 51 + 43 42	OC	7.6	18′	
VIII-43	1750	TAU	05 04 + 23 39	OC	–	–	
VIII-61	1778	AUR	05 08 + 37 03	OC	7.7	7′	
VIII-41	1802	TAU	05 10 + 24 06	NE	–	–	
VIII-4	1896	TAU	05 25 + 20 10	NE	–	–	
VIII-42	1996	TAU	05 38 + 25 49	NE	–	–	
VIII-28	2026	TAU	05 43 + 20 07	NE	–	–	
VIII-2	2063	ORI	05 47 + 08 48	NE	–	–	
VIII-26	2129	GEM	06 01 + 23 18	OC	6.7	7′	
VIII-68	2126	AUR	06 03 + 49 54	OC	10p	6′	
VIII-24	**2169**	**ORI**	**06 08 + 13 57**	**OC**	**5.9**	**7′**	**The "37" Cluster**
VIII-6	2180	ORI	06 10 + 04 43	NE	–	–	
VIII-25	**2232**	**MON**	**06 27 – 04 45**	**OC**	**3.9**	**30′**	
VIII-9	2234	GEM	06 29 + 16 41	NE	–	–	
VIII-49	2240	AUR	06 33 + 35 12	NE	–	–	
VIII-3	2251	MON	06 35 + 08 22	OC	7.3	10′	
VIII-50	2252	MON	06 35 + 05 23	OC	8p	20′	
VIII-48	2260	MON	06 38 – 01 28	NE	–	–	
VIII-5 = V-27	**2264**	**MON**	**06 41 + 09 53**	**OC**	**3.9**	**20′**	**Christmas Tree Cluster**
VIII-31	2286	MON	06 48 – 03 10	OC	7.5	15′	
VIII-71	**2281**	**AUR**	**06 49 + 41 04**	**OC**	**5.4**	**15′**	
VIII-39	2302	MON	06 52 – 07 04	OC	8.9	2′	
VIII-51	2306	MON	06 55 – 07 11	NE	–	–	
VIII-60	2311	MON	06 58 – 04 35	OC	10p	7′	
VIII-1B	2319	MON	07 01 + 03 04	NE	–	–	
VIII-40	2331	GEM	07 07 + 27 21	OC	9p	18′	
VIII-32	2335	MON	07 07 – 10 05	OC	7.2	12	
VIII-33	2343	MON	07 08 – 10 39	OC	6.7	7′	
VIII-34	2353	MON	07 15 – 10 18	OC	7.1	20′	
VIII-45	2358	CMA	07 17 – 17 03	NE	–	–	
VIII-27	2367	CMA	07 20 – 21 56	OC	7.9	4′	
VIII-35	2374	CMA	07 24 – 13 16	OC	8.0	19′	
VIII-44	2394	CMI	07 29 + 07 02	NE	–	–	
VIII-11	2395	GEM	07 27 + 13 35	OC	8.0	12′	
VIII-36	2396	PUP	07 28 – 11 44	OC	7p	10′	

(Continued)

Class VIII: Coarsely Scattered Clusters of Stars (89)—(Continued)							
VIII-52	2413	PUP	07 33 – 13 06	NE	–	–	
VIII-37	2414	PUP	07 33 – 15 27	OC	7.9	4'	
VIII-38	**2422**	**PUP**	**07 37 – 14 30**	**OC**	**4.4**	**30'**	**=M47**
VIII-87	2425	PUP	07 38 – 14 52	OC	–	3'	
VIII-47	2428	PUP	07 39 – 16 31	NE	–	–	
VIII-46	2430	PUP	07 39 – 16 21	NE	–	–	
VIII-1	**2509**	**PUP**	**08 01 – 19 04**	**OC**	**9.3**	**4'**	
VIII-30	2527	PUP	08 05 – 28 10	OC	6.5	22'	
VIII-10	2678	CNC	08 50 + 11 20	NE	–	–	
VIII-53	6507	SGR	18 00 – 17 24	OC	10p	7'	
VIII-54	6561	SGR	18 10 – 16 48	NE	–	–	
VIII-55	6596	SGR	18 18 – 16 40	OC	–	–	
VIII-15	6604	SER	18 18 – 12 14	OC	6.5	2'	
VIII-72	**6633**	**OPH**	**18 28 + 06 34**	**OC**	**4.6**	**27'**	
VIII-14	6647	SGR	18 32 – 17 21	OC/ NE?	8p	–	(See Chapter 13)
VIII-12	**6664**	**SCT**	**18 37 – 08 13**	**OC**	**7.8**	**16'**	
VIII-13	6728	AQL	19 00 – 08 57	NE	–	–	
VIII-81	6793	VUL	19 23 + 22 11	OC	–	–	
VIII-21	6800	VUL	19 27 + 25 08	OC	–	–	
VIII-73	6828	AQL	19 50 + 07 55	NE	–	–	
VIII-16	6834	CYG	19 52 + 29 25	OC	7.8	5'	
VIII-18	6837	AQL	19 54 + 11 41	OC	–	–	
VIII-19	6840	AQL	19 55 + 12 06	OC	–	–	
VIII-86	6874	CYG	20 08 + 38 14	NE	–	–	
VIII-22	**6882**	**VUL**	**20 12 + 26 33**	**OC**	**8.1**	**18'**	
VIII-20	**6885**	**VUL**	**20 12 + 26 29**	**OC**	**8.1**	**20'**	**=C37**
VIII-83	6895	CYG	20 16 + 50 14	NE	–	–	
VIII-56	**6910**	**CYG**	**20 23 + 40 47**	**OC**	**6.7**	**8'**	
VIII-17	6938	VUL	20 35 + 22 15	NE	–	–	
VIII-23	6950	DEL	20 41 + 16 38	NE	–	–	
VIII-82	6989	CYG	20 54 + 45 17	NE	–	–	
VIII-76	6991	CYG	20 57 + 47 25	OC	–	–	
VIII-58	6997	CYG	20 57 + 44 38	OC	10p	15'	
VIII-57	7024	CYG	21 06 + 41 30	NE	–	–	
VIII-74	7031	CYG	21 07 + 50 50	OC	9.1	5'	
VIII-67	7160	CEP	21 54 + 62 36	OC	6.1	7'	
VIII-63	7234	CEP	22 12 + 56 58	NE	–	–	
VIII-75	**7243**	**LAC**	**22 15 + 49 53**	**OC**	**6.4**	**21'**	**=C16**
VIII-77	7380	CEP	22 47 + 58 06	OC	7.2	12'	
VIII-69	7686	AND	23 30 + 49 08	OC	5.6	15'	
VIII-62	7708	CEP	23 34 + 72 55	NE	–	–	

About the Author

The author, shown holding a copy of his book *Celestial Harvest: 300-Plus Showpieces of the Heavens for Telescope Viewing & Contemplation*. Originally self-published in 1998 (and updated in 2000), it was reprinted in 2002 by Dover Publications. More than 40 years in the making, its showpieces include the best of the Herschel objects. Courtesy of Warren Greenwald.

James Mullaney is an astronomy writer, lecturer, and consultant who has published more than 500 articles and five books on observing the wonders of the heavens, and logged over 20,000 hours of stargazing time with the unaided eye, binoculars and telescopes. Formerly Curator of the Buhl Planetarium and Institute of Popular Science in Pittsburgh and more recently Director of the DuPont Planetarium, he served as staff astronomer at the University of Pittsburgh's Allegheny Observatory and as an editor for *Sky & Telescope*, *Astronomy* and *Star & Sky* magazines. One of the contributors to Carl Sagan's award-winning *Cosmos*

The author shown with his 5-inch Celestron Schmidt-Cassegrain optical tube assembly mounted on a sturdy vintage Unitron altazimuth mounting with slow motion controls. With excellent optics and a total weight of just 12 pounds, this highly portable instrument can go anywhere and is a joy to use. Despite its relatively small aperture, it shows all of the Herschel objects described in this book. Photo by Sharon Mullaney.

PBS-Television series, his work has received recognition from such notables (and fellow stargazers) as Sir Arthur Clarke, Johnny Carson, Ray Bradbury, Dr. Wernher von Braun, and former student – NASA scientist/astronaut Dr. Jay Abt. His 50 year mission as a "celestial evangelist" has been to "Celebrate the Universe!" – to get others to look up at the majesty of the night sky and to personally experience the joys of stargazing. In February of 2005 he was elected a Fellow of the prestigious Royal Astronomical Society (London). His previous two books for Springer are *Double and Multiple Stars and How to Observe Them* (2005) and *A Buyer's and User's Guide to Astronomical Telescopes and Binoculars* (2007).

Index

Printed in the United States
By Bookmasters